THE HUMANISTIC VIEW OF MAN

THE
HUMANISTIC
VIEW OF MAN

JAIDEEP SINGH

BLUEROSE PUBLISHERS
India | U.K.

Copyright © Jaideep Singh 2023

All rights reserved by author. No part of this publication may be reproduced, stored in a retrieval system or transmitted in any form or by any means, electronic, mechanical, photocopying, recording or otherwise, without the prior permission of the author. Although every precaution has been taken to verify the accuracy of the information contained herein, the publisher assumes no responsibility for any errors or omissions. No liability is assumed for damages that may result from the use of information contained within.

BlueRose Publishers takes no responsibility for any damages, losses, or liabilities that may arise from the use or misuse of the information, products, or services provided in this publication.

For permissions requests or inquiries regarding this publication,
please contact:

BLUEROSE PUBLISHERS
www.BlueRoseONE.com
info@bluerosepublishers.com
+91 8882 898 898
+4407342408967

ISBN: 978-93-93384-22-5

Typesetting: Pooja Sharma

First Edition: November 1979 by IIPA
Second Edition: October 2023

Dedicated

to

CONSYNADI

FOREWORD

The Humanistic View of Man by Dr. Jaideep Singh is an exploration of the nature of human beings as viewed by contemporary humanistic psychologists. Its main finding is that self-actualization and self-determination are the two most central characteristics of what it means to be human.

An analogy can clarify this finding. It was the propulsion to rule oneself, to develop and contribute to the world community according to its own genius, that led India to struggle for its freedom from British rule. What contemporary humanistic psychologists are saying is that this very same propulsion, the urge to rule oneself, to develop oneself and to contribute according to one's own genius, is the primary nature of each and every human being. This implies that the primary responsibility of each person is to live, evolve and contribute according to the wisdom and genius of his or her own unique self. It is in this way that man can make better contribution to social good and establish harmonious relationship with fellow human beings in all work situations.

The author has made a valuable contribution to the exploration of the nature of man and his psychology. Though slightly unusual in approach, I believe that this volume of Dr. Jaideep Singh will be of use to the social psychologists, public administrators, management experts and informed citizens in general who are interested in the development of a social order based on freedom, human dignity and social justice.

(T.N. CHATURVEDI)

Director

November 1979   **INDIAN INSTITUTE OF**

PUBLIC ADMINISTRATION

ACKNOWLEDGMENTS

Speaking as a gardener who brings forth a flower from a seed, I wish to acknowledge my deep indebtedness to Dorothy L. Harris, whose unlimited trust and continuing guidance have been the priceless sunshine and water; Charles C. Case and W. Ray Rucker, whose remarks of caution have ensured that the bloom got protected from weeds; the United States International University for being the nourishing soil it is; people-at-large, who have been working to create a more humanistic atmosphere in the world; and finally, my thanks to life for the seed itself.

CONTENTS

1

INTRODUCTION

Man's behaviour is greatly influenced by his view of man. Gardner Murphy has expressed it thus: "As man makes new images of himself, he indulges in self-fulfilling prophecies. He has always made himself, into what he imagined himself to be."[1]

PURPOSE AND RATIONALE

The purpose of this study is to develop a synthesis of the views of man that are held by a group of leading, contemporary, humanistically-oriented, behavioral scientists.

Over the past twenty-five years the event of very great significance for the future of mankind has been the development of a more humane conceptualization of the nature of man by scientists in the field of human behavior.

This new view of man considers him to be an active, consciously self-determining subject with the freedom and responsibility to be and become his authentic self.

In an article entitled, "The New Copernican Revolution" Willis Harmon proposes that the impact of the recent trend toward empirical research into man's view of himself, of his subjective assumptions regarding what he is and what he can be, may be "even more far-reaching than those which emerged from the Copernican, Darwinian, and Freudian revolutions."[2]

In speaking of this trend, Abraham Maslow writes:

I must confess that I have come to think of this humanist trend in psychology as a revolution in the truest, oldest sense of the word, the sense in which Galileo, Darwin, Einstein, Freud, and Marx made revolutions, *i.e.*, new ways of perceiving and

---

[1] Gardner Murphy, "The Unfolding Images of Man," *Human Potentialities,* ed. Herbert A. Otto (St. Louis, Missouri : Warren H. Green, Inc., 1968), p. 12.

[2] Willis Harmon, "The New Copernican Revolution," The Journal of Humanistic Psychology, IX, No. 2 (Fall 1969), p 128.

thinking, new images of man and of society, new conceptions of ethics and of values, new directions in which to move.[3]

The present study has been uniquely designed to obtain a more penetrating insight into this contemporary, humanistic view of man.

STATEMENT OF THE PROBLEM

The problem is stated below in the form of three questions:

I.   What can be determined regarding the areas and strength of agreement and disagreement in the views of man held by a group of leading, contemporary, humanistically-oriented, behavioral scientists?

II.   What is the synthesis obtained from an analysis of a selection of contemporary, humanistic views of man?

III.   Given a synthesis of the humanistic view of man what would be the evaluation of a natural scientist who is a recognized authority?

ANALYSIS OF THE PROBLEM

The first question requires a content analysis of selected positions. The purpose is twofold: to find the points of agreement and disagreement; to test the strength of each point.

The second question requires the construction of a synthesis. What are the key areas of common emphasis?

The third question requires an evaluation from a recognized authority who is a natural scientist as opposed to a behavioral scientist. Does the natural scientist view man differently?

DELIMITATIONS

The study will be limited to an analysis of the theoretical positions of five scientists.

The behavioral scientists will be drawn primarily from psychology.

The schema for content analysis has been developed by the researcher.

---

[3] Abraham H. Maslow, *Toward a Psychology of Being* (New York: Van Nostrand Reinhold Company, 1968), p. iii.

ASSUMPTIONS

Given the selection criteria as outlined in the chapter on methodology, an analysis of five theoretical positions is sufficient to establish a synthesis.

The summary of the theoretical position is a valid representation of the original as far as the scientist's view of man is concerned.

Content analysis based upon the designed instrument is a valid procedure for accomplishing the purpose of this research study.

DEFINITION OF TERMS

Humanistic: More human-valuing; an adjective implying emphasis upon the distinctively human aspects of the human animal.

Self: The whole person, an integrated unity of body and personality.

Behavioral scientist: A scientist whose primary concern is the study of the nature of man and his relations with his fellow man.

Natural scientist: A scientist whose primary concern is the study of one or more of the physical and biological sciences.

PREVIEW

The following chapter consists of a review of selected literature to provide a suitable background for this study on the contemporary, humanistic view of man. Chapter 3 discusses the methodology used by this study and presents the content analysis instrument.

The following five chapters will each include a summary and analysis of each of the five theoretical positions selected. Chapter 9 presents the synthesis drawn from the previous analyses. Chapter 10 is a report of the evaluation of this synthesis by a prominent natural scientist and Chapter 11 concludes the study with a summary of the tentative conclusions reached and an appraisal of Bronowski's comments on the dissertation.

2

REVIEW OF RELATED LITERATURE

The review has been divided into five sections. The first section reviews the book, *Humanistic Viewpoints in Psychology* edited by Frank T. Severin. The second section analyzes Corliss Lamont's, *The Philosophy of Humanism* while the third surveys *New Views of the Nature of Man* edited by John R. Platt. The fourth section focuses on a dissertation by William C. Charron entitled : "An Exposition and Analysis of William James's Views on the Nature of Man." The final section presents a theoretical background of the content analysis approach to the analysis of qualitative material. Needless to say, during the reviewing process, the purpose of the study has always been kept sharply in focus.

HUMANISTIC VIEWPOINTS IN PSYCHOLOGY[1]

There are three interwoven conceptualizations in the view of man emphasized in this collection of articles:

1. Man is a holistic totality, a whole greater than the addition of his part traits, processes and functions.
2. Man is self-determining, self-governing.
3. Man is unique.

Referring to the first conceptualization, we note that James F.T. Bugental states this as the defining characteristic of man:

I propose that the defining concept of man basic to the new humanistic movement in psychology is that *man is the process that supersedes the sum of his part functions.*[2]

---

[1] Frank T. Severin (ed.), *Humanistic Viewpoints in Psychology* (New York : McGraw-Hill Book Company, 1965).

[2] James F.T. Bugental, "Humanistic Psychology : A New Breakthrough", *Humanistic Viewpoints in Psychology*, ed. Frank T. Severin (New York : McGraw-Hill Book Company, 1965), p. 9.

Gordon Allport gives a short summary of personalistic formulations on the nature of man. These emphasize his holistic nature.

There are several versions of personalistic thought. They all agree that the individual person as a patterned entity must serve as the centre of gravity for psychology. The intention of personalism is to rewrite the science of mental life entirely around this focus.

....

Without the coordinating concept of person (or some equivalent, such as self or ego), it is impossible to account for the interaction of psychological processes. Memory affects perception, desire influences meaning, meaning determines action, and action shapes memory; and so on indefinitely. This constant interpenetration takes place within some boundary, and the boundary is the person. The flow occurs for some purpose, and the purpose can be stated only in terms of service to the person.

The organization of thought or behavior can have no significance unless viewed as taking place within a definite framework. Psychological states do not organize themselves or lead independent existences. Their arrangement merely constitutes part of a larger arrangement - the personal life. Such concepts as *function, adaptation, use* have no significance without reference to the person...If an adjustment takes place it must be an adjustment *of* something, *to* something, *for* something. Again, the person is central.

All the evidence - introspective and otherwise - that forces psychology to take account of the *self* is here relevant. The very elusiveness of the self - James says that to grasp it fully in consciousness is like trying to step on one's own shadow - proves that it is the ground of all experience. Although seldom salient itself, it provides the platform for all other experience.[3]

Man is self-governing, self-determining, responsible for his being and becoming. This second conceptualization regarding

---

[3] Gordon W. Allport, "The Person in Psychology," *Humanistic Viewpoints in Psychology*, ed. Frank T. Severin (New York . McGraw-Hill Book Company, 1965), pp. 38-39.

the nature of man rests on the first. Maurice Tamerlin explains :

> . . .the relationship between personal responsibility and the experience of free choice is intimate : without a prior feeling of freedom to choose between alternatives, the individual does not view himself as responsible.

A sense of personal responsibility, then, seems to follow the experience of free choice. The sequence raises the question : What experiences precede a feeling of free choice? At the experiential level, choice depends upon an experience of self, or of personal identity. To choose implies an identity, a sense of self or "I," *a person* who chooses in terms of his own thoughts and wishes. Choice requires a chooser ... The point is that the experience of choice requires a sense of self; reciprocally, the exercise of choice is an affirmation of selfhood. "Man is never more human than at the moment of decision," as Tillich put it.[4]

The third conceptualization, *viz.*, man is unique, *i.e.*, each person is like no other person that now exists, ever existed in the past or ever shall exist in the future, is especially stressed by the existentialist. To quote Allport again :

> Each person is busy building his own peculiar constellation of ego-world relationships. His motives are his own, taking always the form of "personal projects." His inheritance is unique; his experienced environment is unique; all his ego-world relationships are unique. Existence ultimately resides nowhere except in the individual's point of view. Certainly no counselor or therapist can succeed unless he can understand the patient's dilemma from the patient's standpoint. A million mortals will experience their ego-world quandaries in a million ways.
>
> Thus, at bottom, the existentialist approach to man is urgently idiographic...[5]

The holistic, self-determining and unique nature of man is summarized well in the following stimulating words of Abraham Maslow :

---

[4] Maurice K. Temerlin, "On Choice and Responsibility in Humanistic Psychotherapy", *Humanistic Viewpoints in Psychology*, ed. Frank T. Severin (New York : McGraw-Hill Book Company, 1965), p. 74

[5] Allport, *op. cit.*, p. 43.

. . .The human being is an irreducible unit, at least as far as psychological research is concerned. Everything in him is related to everything else, in greater or lesser degree.

This has one important consequence. In his essential core, no human being is comparable with any other. Therefore his ideals for himself, his path of growth must also be unique. His goal must arise out of his own nature, and not be chosen by comparison or competition with others. Each man's task is to become the best himself. Joe Doakes must not try to be like Abraham Lincoln or Thomas Jefferson or any other model or hero. He must become the best Joe Doakes in the world. This he can do, and only this is necessary or possible...[6]

THE PHILOSOPHY OF HUMANISM[7]

This section has been presented in the following three parts :

1. The definition of humanism.
2. On the unity of body and personality.
3. Choice, chance and determinism.

THE DEFINITION OF HUMANISM

This subdivision presents a brief yet comprehensive look at the view of man espoused by contemporary Humanism. According to Lamont, there are ten central propositions to the philosophy of Humanism :

First, Humanism believes in a naturalistic metaphysics or attitude toward the universe that considers all forms of the supernatural as myth; and that regards Nature as the totality of being and as a constantly changing system of matter and energy which exists independently of any mind or consciousness.

Second, Humanism, drawing especially upon the laws and facts of science, believes that man is an evolutionary product of

---

[6] Abraham H. Maslow, "A Philosophy of Psychology : The Need for a Mature Science of Human Nature," *Humanistic Viewpoints in Psychology*, ed. Frank T. Severin (New York : McGraw-Hill Book Company, 1965), p. 32.
[7] Corliss Lamont, *The Philosophy of Humanism* (New York : Frederick Ungar Publishing Company, 1965).

the Nature of which he is a part; that his mind is indivisibly conjoined with the functioning of his brain; and that as an inseparable unity of body and personality he can have no conscious survival after death.

Third, Humanism, having its ultimate faith in man, believes, that human beings possess the power or potentiality of solving their own problems, through reliance primarily upon reason and scientific method applied with courage and vision.

Fourth, Humanism, in opposition to all theories of universal determinism, fatalism or predestination, believes that human beings while conditioned by the past, possess genuine freedom of creative choice and action, and are, within certain objective limits, the masters of their own destiny.

Fifth, Humanism believes in an ethics or morality that grounds all human values in this-earthly experiences and relationships and that holds as its highest goal the this-worldly happiness, freedom, and progress - economic, cultural, and ethical - of all mankind, irrespective of nation, race, or religion.

Sixth, Humanism believes that the individual attains the good life by harmoniously combining personal satisfactions and continuous self-development with significant work and other activities that contribute to the welfare of the community.

Seventh, Humanism believes in the widest possible development of art and the awareness of beauty, including the appreciation of Nature's loveliness and splendor, so that the aesthetic experience may become a pervasive reality in the life of man.

Eighth, Humanism believes in a far-reaching social program that stands for the establishment throughout the world of democracy, peace, and a high standard of living on the foundations of a flourishing economic order, both national and international.

Ninth, Humanism believes in the complete social implementation of reason and scientific method; and thereby in the use of democratic procedures, including full freedom of expression and civic liberties, throughout all areas of economic, political, and cultural life.

Tenth, Humanism, in accordance with scientific method, believes in the unending questioning of basic assumptions and convictions, including its own. Humanism is not a new dogma, but is a developing philosophy ever open to experimental testing, newly discovered facts, and more rigorous reasoning.[8]

------

[8] Lamont, *op. cit.*, pp. 12-44.

ON THE UNITY OF BODY AND PERSONALITY

Humanism believes in the view that man is an integrated totality, an indivisible unity of body and personality. Thus, the body and personality exist together or not at all. After the death of a person, he ceases to exist. Men have available to them only this one life - let them make the most of it - there is no possibility of life after death in any form.

Thus, Lamont queries and concludes :

For Humanism, as for most philosophies, the most important and far-reaching problem connected with the nature and destiny of man is what sort of relationship exists between the physical body and the personality, which includes the mind in its every aspect. Is the relation between body and personality so close and fundamental that they constitute an indissoluble unity (the monistic theory); or is it so loose and unessential that the personality may be considered a separate and independent entity which in the final analysis can function without the body (the dualistic theory). . .?[9]

To summarize, my brief survey...builds up a most compelling verdict in support of the unbreakable unity of the body and personality, including the mind and consciousness. Testifying always and everywhere to the union, one and inseparable, between body and personality, the monistic or naturalistic psychology stands today as one of the greatest achievements in the history of science. . .[10]

CHOICE, CHANCE AND DETERMINISM

The Humanist philosophy believes that each of these, choice or free will, chance or luck, and determinism or necessity are interacting aspects of reality. Particularly, it stresses that man has the capacity of freedom of choice, of free will, and that he can, if he so chooses, make a difference in how he lives and what he becomes.

Lamont writes :

Personal freedom of choice ("free will" in traditional terminology) means the capacity of conscious men to make real decisions in situations where significant alternatives exist.

---

[9] Lamont, op cit, p 81
[10] Ibid., p. 94.

Obviously, physical, economic, social, and other factors always condition and limit human choices. Our mental and physical inheritance, the extent and type of our education, our income and kind of job, in fact, our total environment past and present, all influence our current behavior. . .[11]

My analysis ought to have made plain that to phrase the central issue under discussion "freedom of choice *versus* determinism" is quite misleading, because it assumes at the outset that the two concepts are mutually exclusive. On the contrary, there is in human life a constant, interlocking pattern of *both* freedom and determinism. It is not, then, the role of human freedom in general to combat determinism, but to work with it, tame it and employ it for the achievement of worthwhile purposes.

Interacting everywhere with freedom and determinism is contingency, the chance event, the conditional happening. The fact that human freedom is inextricably linked with the pervasive contingency in our world shows that free choice is a perfectly natural phenomenon and not an inexplicable exception to Nature's ways. . .[12]

Every free choice is equivalent to a *free cause*. In short, *you* — a thinking, initiating, choosing agent - can be and frequently are the free cause of your own actions. . .

. . .if men are genuinely free in the way I have indicated, it follows that groups, communities, nations and civilizations — all of which are composed of human individuals — likewise in the large possess freedom of choice.[13]

NEW VIEWS OF THE NATURE OF MAN[14]

Three articles from this collection have been selected as the three subdivisions of this section.

WILLARD F. LIBBY: MAN'S PLACE IN THE PHYSICAL UNIVERSE

Willard Libby, winner of the 1960 Nobel prize in chemistry,

---

[11] Lamont, *op. cit.*, p. 159.

[12] Ibid., p. 168

[13] Ibid., p. 169

[14] John R. Platt (ed.), New Views of the Nature of Man (Chicago : The University of Chicago Press, 1965)

takes a very broad look at man and his environment. He makes three main points :

. . .First, we are living in a part of the universe where there was a beginning and where there must be an end. We are in between and progressing toward oblivion; yet we can foresee that some billions of years may lie ahead.

Second, life is a natural chemical consequence of the environment; so we can expect it to occur in the vicinity of many, many other stars. This is what we have called the Principle of Life. Third, man is new on earth and is a unique form of terrestrial life embodying what we may call the Principle of Intelligence.[15]

With respect to the first point, Libby cautions that he is not postulating the end of the universe as a whole, only that part of it in which we live — the solar system — definitely will run down with the sun no longer shining and the planets falling into the sun. But the time period he mentions is so incredibly huge — billions of years — that it would be quite miraculous if man does not destroy himself before then without this being done through natural processes.

As regards the second point, Libby notes that though life is a natural consequence of our planetary environment and in all probability occurs widely in other stellar systems, no scientist at present has been able to explain the magical transition from the inanimate to the animate and he speculates that it may be centuries before we can make even the simplest form of living matter synthetically.

The third point is particularly emphasized by Libby :

. . .man's place in the physical universe is to be its master or at least to be the master of the part that he inhabits. It is his place, by controlling the natural forces with his intelligence, to put them to work to his purposes and to build a future world in his own image. The possibility of doing this is exciting. It is what can be done if man has the strength to control his irrational tendencies, to suppress his weaker divisionary tendencies, and to develop his strongest and best characteristics. Everything we

---

[15] Willard F. Libby, "Man's Place in the Physical Universe, "*New Views of the Nature of Man,* ed. John R. Platt (Chicago : The University of Chicago Press, 1965), p. 9

know would indicate that the opportunities for future development are unbounded for a rational society operating without war. Man's intelligence, self-respect, sense of responsibility, and feeling of destiny are the qualities that will carry us forward. But for all these hopes to come true, man must enjoy his role as king of the universe. He must understand that this is his function; he must have enough responsibility to carry it out without leading himself toward death and self-destruction. This, to me, is man's place in the physical universe : to be its king through the power he alone possesses - the Principle of Intelligence.[16]

GEORGE WALD: DETERMINACY, INDIVIDUALITY AND THE PROBLEM OF FREE WILL

Wald uses, as a starting point for his comments on determinism, individuality and free will, the Heisenberg's Uncertainty Principle which states that the more accurately one succeeds in measuring the position, at a given instant of time, of any elementary particle such as an electron, proton, neutron or photon, the less certain one is of its velocity, or vice versa. The ultimate uncertainty regarding the behavior of particles is formidable and seems to represent a fundamental property of the nature of the particles and hence of the nature of the universe since matter is composed of them. Thus, we can confidently say that nature, and therefore man, is ultimately unpredictable. Wald mentions in this connection that,

In a discussion which stretched over many years, Einstein and Bohr argued this point, Bohr maintaining and Einstein continuing to doubt that the uncertainty principle expresses ultimate reality. Most physicists agreed with Bohr, and still do.[17]

But, though one cannot define the locations, sizes and shapes of elementary particles, nor even the shapes of atoms, at the level of molecules, however, which are aggregates of atoms, morphology, *i.e.*, a definite form with respect to size and shape,

---

[16] Libby, *op. cit.*, p. 14

[17] George Wald, "Determinacy, Individuality and the Problem of Free Will," *New Views of the Nature of Man,* ed. John R. Platt (Chicago : The University of Chicago Press, 1965), p. 18.

emerges. And living cells being particular molecular constructs, are directly determined by the shapes of their component molecules. Thus, man being a synthesis of cells, is very much a determined organism; and his behavior is determined.

Wald then discusses the phenomena of individuality. He explains :

. . .There are no two living cells, and I would venture to say there never have been two living cells, that are or were identical.
. .

What kind of thing is a living organism to present this extraordinary individuality?

For one thing, living organisms are enormously complex. They are organized associations of great numbers of very different kinds of molecules known to chemistry. . . The enormous complexity of the composition of living organisms in itself makes identity very improbable.

That complexity is compounded by the fact that living cells have not a static but a dynamic composition. They are the loci of a constant inflow and outflow of energy and material. . . Living organisms are individual not only in space but in time. They grow old; they acquire new characters; they bear the scars of experience - and all these things make them recognizably different at every stage in life.

A further reason for the rigorous individuality of living organisms is that the genetic information that determines them - a monkey, an amoeba, a bacterial cell - is laid out in such nucleic acid molecules...in the form of a molecular tape in which the four kinds of nucleotides which are the units of nucleic acid structures are linked in specific sequences in one continuous chain. It is the reading of those sequences that determines the entire eventual structure and composition, and even aspects of the behavior, of living organisms.[18]

. . . .

...through the factors we have been discussing, the complexity of the organism, its dynamic state, and the constant intrusion of genetic disorder, one can be quite convinced that each living thing, including every man, is unique, an individual unlike any other in space or time. But to those factors one must

---

[18] Wald, *op. cit.*, pp. 23-25.

add another of ultimate importance. It is that living organisms store history. Not only does each of us come into the world with a unique composition and inheritance, but to those we begin to accumulate a unique experience. That personal history, growing throughout our lives, is ours alone. That private self that is you or I is the unique composition and structure that come to us via metabolism and inheritance, coupled with a unique personal history that is forever growing.[19]

Finally, Wald refers to the perennial question of how man can have free will, *i.e.*, freedom of choice when only a little earlier in the same article he has been depicted as a completely determined organism. Wald says :

. . .I see no essential incompatibility between such a complete determinism of behavior and free will. Behavior may all be determined, but it is surely not all predictable; and I think that the essence of our free will lies in that unpredictability.

. . .the essence of free will is not a failure of determinism but a failure of predictability.

How free is free will? It is rather curious that, for all the enthusiasm it arouses, it involves so narrow a segment of our experience and choice. A large realm of our behavior is completely determined and predictable; it is that which permits us to go on living as organized and integrated beings. There is, for example, all the interplay and mutual adjustment of our parts that permits them to function together harmoniously, all the internal adjustment that constitutes our vegetative life. None of that calls for decisions; those functions are too important for decision and are safeguarded by being made automatic.[20]

He continues thus :

. . .I would suppose that such forced and automatic activities account for most of our behavior.

At best, therefore, our will is free only within limits. The more one thinks about it, the more one realizes how narrow those limits are.[21]

And he finally concludes by saying that :

In sum, I think that we have freedom of will and that it

<hr>

[19] Wald, *op. cit.*, p. 33.
[20] *Ibid.*, pp. 36-37.
[21] *Ibid.*, p. 39

comes out of our uniqueness as individuals, perhaps wholly determined, yet to some degree unpredictable. However limited in scope, it is one of our most precious possessions. As such we should seek to enlarge it. . .[22]

ROGER W. SPERRY: MIND, BRAIN, AND HUMANIST VALUES

In this article, Sperry criticizes the materialistic approach that the brain-behavior sciences have espoused over the past half-century. He considers himself among the 0.1 per cent or so minority group of brain researchers who hold that mental phenomena like mind, consciousness, free will, etc., are not illusions and ignorable epiphenomena but the most central, the commanding-governing realities. This latter approach - the mentalist view of man - puts mind over matter and is diametrically opposite to the materialistic view of man. Sperry makes clear that the scientific facts necessary to decide conclusively which of these two approaches is correct just do not exist at present. He writes :

. . .I think we must all agree that neither is going to win the match on the basis of direct, factual evidence. The facts simply do not go far enough to provide the answer, or even to come close. Those centermost processes of the brain with which consciousness is presumably associated are simply not understood. They are so far beyond our comprehension at present that no one I know of has been able even to imagine their nature...[23]

It is Sperry's opinion, however, that the mentalistic view of man, which regards mind and consciousness as the central causal agents that direct the brain's physiological-physical-chemical processes more than vice versa, is the correct model.

He comments :

. . .It is a scheme that idealizes ideas and ideals over physical and chemical interactions, nerve impulse traffic, and DNA. It is a brain model in which conscious mental psychic forces are recognized to be the crowning achievement of some five

---

[22] Wald, *op. cit.*, p. 46

[23] Roger Sperry, "Mind, Brain, and Humanist Values," *New Views of the Nature of Man*, ed. John R. Platt (Chicago: The University of Chicago Press, 1965), p. 76.

hundred million years or more of evolution.[24]

Sperry then attacks a second enemy of the humanistic view of man, *viz.*, the prevailing, scientific rejection of free will. According to Sperry, man has a considerable degree of free will provided this is interpreted to mean self-determination. He gives his opinion thus :

. . .It should be clear by now that in the brain model described here, man is provided in large measure with the mental forces and the mental ability to determine his own actions. This scheme thus allows a high degree of freedom from outside forces as well as mastery over the inner cellular, molecular, and atomic aspects of brain activity. Depending on the state of one's will power, the model also allows considerable freedom from lower-level natural impulses and even from occasional thoughts, beliefs, and the like, though not, of course, from the whole complex. In other words, the kind of brain visualized here does indeed give man plenty of free will, provided we think of free will as self-determination. To a very real and large extent, a person does determine with his own mind what he is going to do from among a large number of possibilities. This does not mean, however, that he is free from the forces of his own decision-making machinery. In particular, what this present model does not do is to free a person from the combined effects of his own thoughts, his own reasoning, his own feeling, his own beliefs, ideals, and hopes, nor does it free him from his inherited makeup or his lifetime memories. All these and more, including, yes, unconscious desires, exert in the brain their due causal influence upon any mental decision, and the combined resultant determines the inevitable but self-determined and highly special and highly personal outcome. . .[25]

Finally, Sperry appraises the materialistic-behavioristic view that man's brain gets its start in fetal life as a blank slate and is subsequently conditioned by the environment. He reminds us that :

In this doctrine the mind, or psyche, was believed to develop gradually out of a lifelong chain of successive conditioned-reflex associations, starting in the infant from a few elementary

---

[24] Sperry, *op. cit.*, p. 78.
[25] *Ibid.*, p. 87

reactions, like love and hate, fear and anger. The whole idea of the genetic inheritance of behavior patterns was forcibly renounced, until the term "instinct" became highly discredited in professional circles, its defamation almost equalling that of consciousness. . .[26]

Sperry then adds :

Much of the basic scientific thinking and evidence behind this view has since suffered a series of severe upsets, leading to a current stand that is almost diametrically opposite to the earlier doctrines...The conditioned response, along with other forms of learning, continues to be recognized, of course, as a highly powerful influence, especially in man, but only within limits much narrower than previously supposed.

Within the specialized fields of scientific inquiry involved here, the pendulum of opinion continues at this date to swing in the direction of inheritance. . .[27]

WILLIAM JAMES'S VIEWS ON MAN[28]

James's views on the nature of man have been presented as occurring in five stages as under :

Stage No.	Time Period	View of Man
1.	Early 1870's and 1890's	Man is an adaptive animal
2.	Mid 1890's	Man is a moral hero
3.	Late 1890's	Man is a witness of divinity
4.	1904-1905	Man is a purely physical phenomenon
5.	1906-1910	Man is divinity

Each of these views is discussed below.

---

[26] Sperry, *op, cit.*, p. 89.

[27] *Ibid.*, p. 90.

[28] William C. Charron, "An Exposition and Analysis of William James's Views on the Nature of Man" (Unpublished Doctor's dissertation Marquette University, 1966).

Man Is An Adaptive Animal

This was James's original view. Essentially it states that man is an animal who, having arisen through the process of natural evolution involving the survival of the fittest, strives to survive in the environment by adapting to it.

To quote Charron :

. . .James argues that, functionally considered, the essence of mental life and physical life are the same - both are primarily and properly ordered to the practical end of securing the preservation and well-being of the physical organism. . .[29]

Man Is A Moral Hero

During the mid-1890's, James began to focus on man as a being that lived for the sake of serving ideals— such as honor, valor, truth, integrity, the humanization of man, etc. In the portrayal, heroism and self-sacrifice in the service of an ideal world were the ingredients that specifically characterized man. Man was viewed as the being committed to high ideals and values; as one who was willing to persevere, suffer pain and martyrdom on their behalf.

Thus, Charron explains :

For James, life appears to have a meaning for us only as long as it seems to be for the sake of something higher. No mode of living can ultimately be appreciated by man as an end in itself. As a result, life or any form of vital activity feels insignificant or meaningless when taken by itself. However, by assuming the heroic and strenuous mood, a man feels that his life has meaning because he feels he is living for the sake of some higher purpose. From this perspective, it is obvious that higher ideals are important to James only because commitment to them makes possible the release of man's energy for the heroic life. . .[30]

And he continues by adding :

. . .Zest for life is the mental state characteristic of those in the thick of a struggle for ideals; their blood is up, and their interest is keen. In short, the constitution of human nature is such that no matter what of a man's desires are fulfilled, if his propensites for life *"in extremis"* are not being utilized, he will

---

[29] Charron, *op. cit.*, P. 29.
[30] *Ibid.*, p. 45

eventually slip into deep depression. . .[31]

....

What is needed, then, to actuate a man's propensities for the ennobling type of strenuousness is an adequately stimulating ideal. . .[32]

....

For both James and Nietzsche, life is more becoming than being. The full intensity of life is realized in self-assertion and sacrifice, challenge and daring, not compromise and contentment, tranquility and security. It is only through his power to suffer for his ideals that a man senses most deeply the meaning and dignity of his existence as a living being. . .For man, a perfectly painless life would be a subtle version of hell. He needs struggle and pain in his life in order to really feel he has life and to feel the dignity of that life. . .

For both these philosophers, James and Nietzsche, the great man is the man who distinguishes himself on the battlefield of life with an undaunted determination to fight for his ideals in the face of difficulties and absurdities that lay his weaker brethren low. The weaker ones, the hedonistic comfort-seekers and the stoical preachers of acquiescence. fail to meet the demands of human existence. The purpose of the human drama, for man, is identically the great effort he can make and the personal power he can manifest in the pursuit of the ideals he sets for himself. The extent to which a man makes the heroic effort in struggling for his ideals is the direct measure of that man's worth as a man. . .[33]

MAN IS A WITNESS OF DIVINITY

In the late 1890's, his probing into the questions of human immortality and phenomena like mysticism and spiritualism, led James to believe in a new view of man. This view defines man in terms of an extra-physical and divine dimension of reality. That is, the conscious mind came to be viewed by him as only a small portion of man's total mind - the larger portion being his subconscious.

---

[31] Charron, *op. cit.,* p. 48
[32] *Ibid.,* p. 51.
[33] *Ibid.,* pp. 66-67.

Thus, Charron notes :

. . .James shifts the center of man's being from his physical organism, and even from his conscious mind, to a subliminal or subconscious mind of man. This subliminal region of the mind allegedly constitutes the "larger" and "more important" part of man. This subliminal region is larger in that it is open to dimensions of reality that far exceed that dimension in which we live as physical beings and of which we are conscious as "normal" waking consciousnesses. This same subliminal region is more important to us in that the dimension of reality to which it is open is the divine order of reality. James contends that, from this divine order through his subliminal mind, a man receives the same "saving energies" that actuate saints, mystics, and other religious giants. . .[34]

In contrast to his earlier view, therefore :

. . .James begins to minimize the role of positive volitional effort in the life of man. In relation to the larger subliminal self, James thinks that the waking self is an inferior self which attains its perfection by surrendering itself to the greater power of the subliminal order. Previously, in the middle of the 1890's, James had thought that the only gospel of salvation for man was the gospel of strenuous and heroic will effort. . .

. . .He now argues that the way to the meaningful life is through a surrender to the deeper subliminal powers. . .[35]

MAN IS A PURELY PHYSICAL PHENOMENON

Beginning in 1904, James made a radical change in his most basic assumption. So far he had developed his theories within a dualistic framework, i.e., man consists of an interacting mind and body. Now, James began to argue that there was no such entity as mind - that man is only a body. Thus, Charron notes :

. . .James denies that there is any experience, even introspective experience, of anything that is immaterial, be it a soul, an inner spiritual life, a non-physical stream of consciousness, or what have you. Whether one considers another man or reflects unto himself, the only thing that is experienced is the body and various bodily changes that can be

---

[34] Charron, *op. cit.*, p. 71.
[35] *Ibid.*, pp. 99-100.

mistakenly interpreted as immaterial or spiritual events when in fact they are but felt physiological adjustments, strains, and tensions...all the basic facts of human behavior can be explained without recourse to anything more than the body. All human endeavor is explainable in terms of the physical. . .[36]

MAN IS DIVINITY

This was James's view of man during the last few years before his death in 1910.

Once again, he focuses on man as an interacting synthesis of mind and body, but considers the mind more important than the body.

. . .The human body was relegated to the status of a "weight" which dragged the mind's attention into the physical order. That part of man which is called the waking consciousness was claimed, by James, to be continuous with a wider, but hidden, conscious life. This wider, hidden consciousness, the so-called "subliminal consciousness", was held to be the source of many of the energies experienced by the waking self of man. . .

Again, ...James identifies man with the mind. And again he claims that the normal waking consciousness of man is continuous with a wider mind from which saving experiences come. However, James now goes further... He now claims that the wider self, or wider consciousness, of each man is the one God. Before, in his religious writing, James had been content to imagine that the wider subliminal self of each man was separate and distinct from every other man's; and he had been satisfied to picture the deity (or deities) as transcendent to man and as the environment and object(s) of each man's wider consciousness ... However, in 1908, . . .James's pantheism imagines God on the subject side of human consciousness, as a *co-witness* who is continuous with that human consciousness. As continuous with the larger consciousness which is God, the human consciousness can be viewed as an "internal part" of God. . .[37]

---

[36] Charron, *op. cit.*, pp. 112-113.
[37] *Ibid.*, pp. 147-149.

CONTENT ANALYSIS

The following explanation of the content analysis approach to the analysis of qualitative material is taken from an article by Dorwin P. Cartwright.[38]

The process of classifying qualitative material into appropriate categories so as to describe it in an orderly way is known as "content analysis."

For the purposes of objectivity, the categories or dimensions have to be explicitly specified. For each dimension, an operational definition has to be established so that it is clear what features of the content are to be taken as indication that it falls in one category rather than another. Furthermore :

In drawing up such an operational definition, it is important to begin by designating the units of analysis that are to be used. There are basically two kinds of units to be specified. The first of these may be called the "recording unit", which is the specific segment of the content that is characterized by placing it in a given category. The second kind or unit is the "context unit," which is "the largest body of content that may be examined in characterizing a recording unit." The coder might, for example, count each emotionally loaded word as a recording unit, but refer to an entire paragraph to be sure that he records its correct meaning.[39]

In the content analysis used in this study, the recording unit is the idea expressed by the dimension itself and the context unit is a paragraph. Cartwright mentions one other unit :

The quantitative treatment of symbolic material requires that one specify clearly the unit in terms of which quantification is performed. We shall refer to this as the *unit of enumeration*. . .

...it is customary to take a single respondent as the unit of enumeration. . .[40]

As far as the significance of the content analysis is concerned, it should be apparent that this will depend greatly on the quality of the a priori conception of the dimensions

---

[38] Dorwin P. Cartwright, "Analysis of Qualitative Material", *Research Methods in the Behavioral Sciences*, eds. Leon Festinger and Daniel Katz (New York : Holt, Rinehart and Winston, 1953).

[39] *Ibid.*, p. 437.

[40] *Ibid.*, pp. 440-441.

involved. It is also important in this connection to ensure that the dimensions meet the requirements of logical correctness. A system of categories meets this requirement if it is exhaustive and if its categories are mutually exclusive.

SUMMARY

This chapter has reviewed four selections from the literature relating to the contemporary humanistic view of man. The first selection, *Humanistic Viewpoints in Psychology* edited by Frank T. Severin emphasizes that man is holistic/an integrated whole, self-determining and unique. The second, Corliss Lamont's, *The Philosophy of Humanism,* defines Humanism, strongly supports the view that man is a holistic unity of body and personality, and concludes by pointing out that though choice, chance and determinism are all parts of reality, man's freedom to choose can be the determining factor in how he lives and what he makes of himself. The third selection, *New Views of the Nature of Man* edited by John R. Platt, distinguishes man as the user of intelligence, stresses his unique individuality and his naturally evolved distinctive ability to be his own creator through his ideas and ideals. Finally, in a dissertation entitled : "An Exposition and Analysis of William James's Views on the Nature of Man," William C. Charron shows how James's views on man changed over his lifetime.

A note on the theoretical background of the content analysis approach used in this study forms the concluding section of this chapter.

3

METHODOLOGY

The basic method of the study is to design a suitable instrument for content analysis and to use this on summarized versions of the views of man expressed by five selected, humanistically-oriented scientists. This is done in order to obtain an evaluation of the areas and strength of agreement and disagreement. From this analysis, a synthesis of the views is developed. Finally, comments on the synthesis are to be invited from a leading natural scientist such as Jacob Bronowski.

SELECTION OF THE THEORETICAL POSITIONS

The following criteria have been used for the selection of the behavioral scientists whose theoretical positions will be analyzed :
1. That the scientists be contemporary.
2. That they be leading experts in their field.
3. That their orientation be humanistic.
4. That they have primarily worked in the area of psychology.
5. That their theoretical position is sufficiently delineated and complete to enable analysis.
6. That each scientist, on preliminary consideration, has a position that is adequately distinct from that of the others.

The following five social scientists meet the criteria :
1. Hubert Bonner.
2. James F.T. Bugental.
3. Erich Fromm.
4. Abraham H. Maslow.
5. Rollo May.

DATA ANALYSIS

For data analysis, a Content Analysis Instrument has been designed as a result of a pilot study. The Instrument has been included in this chapter while the pilot study, for the sake of clarity, has been presented as an appendix to the dissertation.

PROCEDURE

1. Complete reading of the theoretical position of the first behavioral scientist.
2. Development of the summarized version of the view of man expressed by the scientist without loss of content significant for this study.
3. Repetition of the above two steps for each of the other four theoretical positions.
4. Analysis and coding of the five summarized views of man using the Content Analysis Instrument.
5. Comparison of the areas and strength of agreement and disagreement among the five views of man.
6. Combination of the five views into a synthesis.
7. Obtain an evaluation of this synthesis from a prominent natural scientist such as Jacob Bronowski.

THE CONTENT ANALYSIS INSTRUMENT

The content Analysis Instrument used in the present study is presented in Table 1. It incorporates a system of dimensions that was considered adequate for analyzing the views of man held by contemporary humanistic psychologists. For purposes of objectivity, each dimension has been operationally defined so as to make clear what features of the content are to be taken as indicative that it comes under one dimension rather than another.

The procedure for the use of the instrument is given below :

1. The summarized version of the view of man is carefully analyzed, page by page, and every time the idea expressed by a particular dimension is observed to occur a count is recorded for that dimension.
2. These counts are added to give an individual total for each dimension.
3. All the individual total are then added to give a cumulative total.
4. The percentage degree of emphasis for each dimension is obtained by using the formula :

$$\frac{\text{Individual total for each dimension}}{\text{Cumulative total for all dimensions}} \times 100$$

25

TABLE 1 THE CONTENT ANALYSIS INSTRUMENT

No.	Dimension/Category	Count	Total	Percentage Degree of Emphasis
1.	A communicative being			
2.	Accepting of non-hedonic emotions			
3.	Always in process			
4.	Conscious/Aware			
5.	Creative			
6.	Forward-thrusting			
7.	Freedom-cherishing			
8.	Future-imagining			
9.	Goal-directed			
10.	Holistic/An integrated whole			
11.	Intelligent			
12.	Much like fellow man			
13.	Nature-appreciating			
14.	Ontologically responsible			
15.	Open to life/experience			
16.	Other-affirming			
17.	Pleasure-loving			
18.	Present-confronting			
19.	Rational			
20.	Risk-taking/Courageous			
21.	Self-active			
22.	Self-actualizing/Self-realizing			
23.	Self-affirming			
24.	Self-determining			
25.	Self-disciplining			
26.	Sexual			
27.	Socially responsible			
28.	Ultimately unknowable			
29.	Unique			
30[a].	Reality-accepting			
30[b].	Ultimately alone			
		Cumulative Total	100%	

[a]The dimensions can be understood as a series of predicates for the sentence beginning, "Man is....."

[b]The dimensions have been listed in alphabetical order apart from the last two dimensions, *i.e.*, 30[a] and 30[b]. These were chosen after a preliminary reading of the five theoretical positions and replaced category No. 30 entitled, 'Other/Miscellaneous.'

One important assumption that underlies this content analysis procedure needs to be noted: the number of counts recorded for a particular dimension is considered to be a valid measure of the emphasis that the scientist himself gives to the idea expressed by that dimension in his view of man.

Finally, it may be observed that the qualitative material in the summaries that were content analyzed is composed of direct quotations from the works of the individual scientists rather than statements interpreting their views of man. This has been done in order to keep the amount of personal interpretation by the researcher at a minimum and hence to maximize the reliability of this research study.

DEFINITION OF TERMS USED IN THE CONTENT ANALYSIS INSTRUMENT

1. *A communicative being:* A relating, transmitting and receiving, sharing and exchanging being.
2. *Accepting of non-hedonic emotions :* Accepting and appreciative of the reality, unavoidability and value of affective experiences like pain, anger, fear, resentment, guilt, anxiety, boredom, loneliness, emptiness, weakness, etc.
3. *Always in process:* Dynamic, ceaselessly changing, continuously becoming.
4. *Conscious/Aware:* Aware of what is, of nature, but more specifically of himself (his likes and dislikes, his strengths and limitations, and his practically infinite potentialities), of his follow man, and the inter-relationship and inter-dependence between them.
5. *Creative:* Explorative, constructive, productive, etc., of phenomena other than himself and his fellow man. The concept of creating his fellow man is included under the dimension 'other-affirming.'
6. *Forward-thrusting:* Proactive; initiative taking, constructively aggressive.
7. *Freedom-cherishing:* Desirous of safe-guarding his freedom of choice, his right to be himself, and of enlarging his margin of freedom so as to be able to become all he can become.
8. *Future-imagining:*An aspiring being; a personification of hope and faith, a dreamer of dreams.
9. *Goal-directed:* Intentional, purposive; an objective setting and accomplishing being.

10. *Holistic/An integrated whole* :An organized totality of body and personality; a unity.
11. *Intelligent* : Able to solve problems, meet life's challenges, live fully through using his conscious and unconscious mind.

The use of processes like intuition, imagination, etc., are included in this category.

12. *Much like fellow man* : Similar to other human beings in the world.
13. *Nature-appreciating:* Appreciative and admiring of the beauty in nature, in the phenomena of life that pervades this infinite, eternal cosmos.
14. *Ontologically responsible:* Responsible for his own self-actualization.
15. *Open to life/experience* : Non-defensively aware, free to experience the world within him and outside of him.
16. *Other-affirming* : Loving and respectful of the other person; facilitative of his self-actualization.
17. *Pleasure-loving* : Desirous of feelings of joy and happiness.
18. *Present-confronting* : Fully absorbed in living in the moment, in the here and now; intensely involved in whatever he does.
19. *Rational* : Logical; sane; sensible; a user of reason.
20. Risk-taking/Courageous : Daring enough to be his authentic self; adventurous.
21. *Self-active* : Dependent on himself for motivating himself, for energizing himself, for propelling himself to be and become his full self.
22. *Self-actualizing/Self-realizing:* Being and becoming his authentic self; utilizing his practically infinite potentialities; becoming what he potentially is. Creative of himself.
23. Self-affirming : Respectful and loving of himself, of his unique nature; authentic; naturally spontaneous.
24. *Self-determining* : Self-defining and self-directing; self-choosing; living rather than being lived.
25. *Self-disciplining* : Self-regulating, self-controlling.
26. *Sexual* : Biologically designed to desire intercourse with a member of the opposite sex.
27. *Socially responsible* : Responsible for facilitating the self-actualization of his fellow men.
28. *Ultimately unknowable* : A mystery, despite all we know or can know about him; always beyond complete understanding.

29. *Unique* : Distinct and irreplaceable; each man is like no other person that presently exists, has ever existed or shall ever exist.
30. *Other/Miscellaneous* : This category is to be used to list any dimensions necessitated by the content analysis that cannot be reasonably accommodated in the other categories mentioned above. The following two dimensions were chosen for this category after a preliminary reading of the five theoretical positions :
30*a*. Reality-accepting.
30*b*. Ultimately alone.

SUMMARY

This chapter has described the methodology used in this study. Essentially, this included the preparing of summarized versions of the views of man held by five contemporary, humanistic, behavioral scientists and the analysis of these summarized views using a specially designed content analysis instrument. The instrument has been presented in this chapter.

The final aspects of the methodology are the comparison of the analyses, their subsequent combination into a synthesis, and the evaluation of this synthesis by a prominent natural scientist.

4

HUBERT BONNER'S VIEW OF MAN

Bonner has given his view of man in his book, *On Being Mindful of Man* and relevant passages from this have been quoted below to form a summarized version of his view.

On Being Mindful of Man expounds my belief that man is a unique, open, and creative individual. The psychological processes of intentionality and proaction, and the uniqueness of individual human behavior, together compose the *Leitmotiv* of my thinking about psychological man . . . That study has also confirmed my beliefs that man is not merely a machine but an integral being, not merely a reality but a potentiality; not merely an ordinary creature, but a superior being.

The view of man that emerges here may be interpreted as a blend of two ancient conceptions, namely, the Hebraic idea of responsibility or duty, and the Hellenic model of excellence or the full actualization of human potentialities. The first stresses a life shared; the second, a life individually perfected. The first is a life of discipline and self-control; the second a life of aspiration and freedom.[1]

The following comments on the nature of man are based upon an analysis of the existential human predicament, i.e., the interactive merging of the human condition and the cultural situation:

---

[1] Hubert Bonner, *On Being Mindful of Man* (Boston : Houghton Mifflin Company, 1965), p. xi.

...although the principle of determinism operates widely, it is not a universal force in the affairs of men. A limited indeterminism is a fact of physical nature and of human life. Although life is governed by innumerable contingencies, it has a wide area of free choice ...

However, modern man is moved far less by his capacity to choose than by his awareness of his own finitude...[2]

....

Anguish, anxiety, and despair are ontological conditions. They are immanent in human nature. Psychotherapy, therefore, cannot eradicate them, and if it did, it would destroy man's human nature. Psychotherapy can perform the supreme task of helping man to understand them as guarantors of his own individuality...[3]

....

...We believe that the separation of man from nature, man from other men and from himself, is fundamentally the convergent product of the human condition and the social situation...[4]

....

...Modern man, who is more mobile than man was in any prior period of history, has only incidental contacts with other men... His separation from his fellow men is a combination of abstract concern and indifference: these are the anonymous and impersonal qualities of the alienated man.[5]

....

...The grave violation of man's moral conscience is not a psychological state merely; not the easily understood guilt-feelings, but a profound condition of human nature. It is ontological, and cannot be transcended by means of psychotherapy or religious redemption. It is a part of human nature... Thus man is guilty when he compromises truth, when he does not combat injustice, when he submits to evil, when he fails to actualize his potentialities...

...We are guilty because we can choose. Guilt, like choice, is an attribute of our human nature, a potentiality of human life.[6]

---

[2] Bonner, *op. cit.*, pp. 71-72.

[3] *Ibid.*, p. 74.

[4] *Ibid.*, p. 77.

[5] *Ibid.*, pp. 77-78.

[6] *Ibid.*, p. 79

Bonner moves on to comment on man's freedom to choose :

...When I choose, I incur an awesome responsibility not only toward myself but toward others as well; for when I choose I make a decision for all those in the orbit of my own behavior and experience... Human beings thus share a common burden, the *awe-full* responsibility of sealing their fate in the act of choice...[7]

Any psychology that probes beneath the overt or public behavior of human beings is inevitably faced by the incontrovertible fact that they are profoundly motivated by their inner experiences. The subjective life has not been adequately described by the Freudian concept of the unconscious. On the contrary, the fully human quality of this inner life is not unconscious but conscious. It consists in the capacity and the act of choice...[8]

Experience thus persuades us to believe in the reality of self-determination, in the capacity of each of us to be an effective individual. Wisdom, if it means anything psychologically, is the recognition by each of us, that he is responsible for his own destiny...[9]

...All living stuff acts toward the consummation of a purpose. If its efforts are frustrated, if its goals are blocked. the organism will reach for the same end by some other route.

In the case of man this principle of goal-direction is normally a deliberate and conscious one. It is an attribute of the human self. Indeed, the mature self is the organized totality of self-regulations, of goal-directedness, of intentions in the process of actualization. This intentionality is an act of free choice among alternate goals, and an act of deliberation concerning the best means of reaching them...[10]

---

[7] Bonner, *op. cit.*, p. 81.

[8] *Ibid.*, p. 82.

[9] *Ibid.*, p. 84.

[10] *Ibid.*, p. 85.

....

Man differs from all other animals in his capacity of choice. In choice, also, lies his individuality and uniqueness. Modern geneticists are agreed that every human being is unique...[11]

....

...Moral behavior is thus one of the highest, if not the supreme, manifestations of volition, deliberation, and choice.

...For more than a century the prevailing psychologies have ignored the problem of moral behavior. This neglect is the natural consequence of a view of man which admitted only the sole or the combined influence of heredity or environment. Man was conceived to be the product of one or both factors, especially in early childhood experience. He could not modify his behavior in the light of his own purposes. He could not choose to transform himself in accordance with his image of himself as a perfectible human being...[12]

....

...at no other period in history has man been more confronted with the inescapable fate of all men: the fact that all men must choose...[13]

....

...No man in possession of his powers can escape accountability for his acts. He alone among animals, has the dreadful responsibility of choosing between good and evil, and he alone is conscious of possessing it...[14]

....

...For all the emphasis on the externality of behavior in recent psychology, for all the panegyrics on the merits of social responsibility, man is first, not last, a self-affirming being. The center of his existence is neither nature nor society, but himself...[15]

....

...One proves his existence, or validates his individuality, by means of the courage to be himself. This self-validation through actualization of one's being, is a matter of moral integrity...[16]

---

[11] Bonner, *op. cit.*, p. 86.
[12] *Ibid.*, p. 88.
[13] *Ibid.*, p. 90.
[14] *Ibid.*, p. 98.
[15] *Ibid.*, p. 103.
[16] *Ibid.*, p. 104.

....

...in the final analysis, the one important fact about the human being is that he is potentially the author of his own history...[17]

....

...no existence, and most particularly no personal existence, is possible in the absence of man's regard for and commitments to, other human beings...[18]

....

...Freedom to choose, ...always incurs responsibility toward others, not only oneself. In the absence of this responsibility, freedom is exploitation...the person whose rights are protected by a free society is morally obligated to guard and enhance the freedom which sustains his own independence...[19]

....

...Contrary to the universal American philosophy that every individual fulfills his purpose in the attainment of happiness, the men who have left a mark upon their fellow beings, have been individuals with the passion to leave the world different from what it was before they were thrust into it ...

Being thus freed from the compulsive search for happiness, the proactive individual can lend himself freely in the service of others. Freedom without commitment is licentiousness...[20]

....

...The reality of the self is validated by its participation in the world of other selves. *Ich und du* - I and thou - together form the continuum of human existence. Apart from another, I am only an object, never a complete person. The alienation of which existentialists have written so movingly, is not due solely to the partition of the individual, but to his separation from other individuals...[21]

....

...Yet, all the conditions of contemporary experience point unequivocally to the fact that man's most desperate longing is not group-relatedness, but self-fulfillment. Man does not seek merely the happiness that a group can provide, but more

---

[17] Bonner, *op. cit.*, p. 106.

[18] *Ibid.*

[19] *Ibid.*, p. 108.

[20] *Ibid.*, p. 110.

[21] *Ibid.*, p. 111.

profoundly the dedicated search for himself. Man is truly his own foundation...[22]

Bonner then focuses on the future-oriented, self-transforming nature of man.

All living things, and most profoundly human beings, complete themselves in the future...[23]

...Man, we have said, is not a passive reactor to stimuli or situations. He is a seeker of future ends. He is not fixated on a single temporal dimension, but unites all of them in himself. His style of life is, nevertheless, an expression of a dynamic forward thrust...[24]

...Psychology must widen its horizon to permit us to view man as a self-directing, freely-choosing, value-creating individual...[25]

...The great moments in the history of human psychology, from Plato to the contemporary growth theories of human personality, have described human nature as always in the process of becoming. The view of man as a proactive being is nurtured in the belief in that which is not yet. It believes that, in a real way, what gives every person his being, his personality, is the person himself. Each creates his reality, his being, in accordance with his vision of who he wants to be... Each of us is a being in the process of becoming...[26]

...Proactive or humanistic psychology is finding increasing evidence in support of the view that man is not wholly - or even largely - a seeker of stability and quiescence...[27]

...Insofar as existence, or being, consists in its becoming, becoming is like the future dimension itself, the primary phenomenon of nature, both material and human...[28]

---

[22] Bonner, *op. cit.*, p. 114.

[23] *Ibid.*, p. 116.

[24] *Ibid.*, p. 121.

[25] *Ibid.*, p. 128.

[26] *Ibid.*, p. 132.

[27] *Ibid.*, p. 135.

[28] *Ibid.*, p. 138.

....

...man desires not only safety and security, but the exultation that comes from adventure and the search for novelty. He has been known to abandon security for the risk of achieving greater fulfillment and a higher level of self-integration...[29]

....

A fascinating aspect of this becoming is that the person who *cares* for the becoming of another person is himself swept along by the other' s self-actualization... Each is directed forward by the becoming of the other. It is this mutuality as well as the sacrificial nature of creative human relationships, that lends to self-affirmation its other-mindedness.[30]

....

...In proactive psychology, ...man is seen as the seeker after values which he sets up himself. From this point of view, more important than tranquility, security, and survival is the individual's desire to fulfill himself as a unique person...[31]

....

...Proactive life, which is to say the truly human life, is indeed an endless becoming...[32]

In conclusion, Bonner writes :

The controlling purpose in this book, which we have sedulously and singlemindedly pursued, is to lay bare the individuality of the human being; to show that his being is constantly changing in the direction implicit in his style of life. Man is a multiform being, seeking to actualize his potentialities. He cannot be meaningfully described by means of such separate elements as drives, motives, memories, and cognitive structures. Rather, he must be viewed as a life-totality.[33]

....

...man is that species of animal who strives to attain a higher

---

[29] Bonner, *op. cit.,* p. 143.

[30] *Ibid.,* p. 143.

[31] *Ibid.,* p. 145.

[32] *Ibid.,* p. 179.

[33] *Ibid.,* p. 181.

state for himself and his fellow human beings through his own efforts...[34]

....

...The conscious self is thus not an epiphenomenon, not the fringe of experience, not a set of habits, but the core of organized and meaningful experience, the organizing and self- regulating capacities of the total personality.[35]

....

The stress on the wholeness of the human being is a marked characteristic of the newer trend in personality theory. Whether we call it organismic, holistic, personalistic, or proactive, the stress is always the same: personality is a relatively consistent and unique whole...[36]

....

The recurrent theme of this book is that man is a creative and proactive being...

...Contrary to the belief that a proactive and humanistic psychology is too easy and optimistic, all the evidence that we can muster shows that self-transformation is the most difficult of all human tasks...[37]

....

Having said all this, we are nevertheless driven to the necessary conclusion that a holistic knowledge of man is an ideal, not a reality. A total knowledge of the whole man is impossible, and we must rest satisfied with partial insights into the whole. In this fact lies both the agony and hope of every sincere investigation of human behavior...[38]

---

[34] Bonnmer, *op. cit.*, pp. 190-191.
[35] *Ibid.*, p. 192.
[36] *Ibid.*, p. 193.
[37]*Ibid.*, p. 226.
[38] *Ibid.*, p. 234.

TABLE 1 BONNER'S VIEW OF MAN : CONTENT ANALYSIS

No.	Page 30	Page 31	Page 32	Page 33	Page 34	Page 35	Page 36	Page 37	Total
1.									0
2.		2							2
3.						3	4		7
4.		1					1		2
5.	1							1	2
6.	1					2	1	1	5
7.	1								1
8.						2			2
9.	1		3						4
10.							4		4
11.									0
12.									0
13.									0
14.		1	1	1					3
15.	1					1			2
16.					1		3		4
17.									0
18.									0
19.									0
20.				1		1			2
21.									0
22.	1			1	1	4	6	1	14
23.				2	1		1		4
24.	1	2	4	4	2	3	1		17
25.	1		1				1		3
26.									0
27.	1		1		3				5
28.								1	1
29.	2			2			2		6
30a.									0
30b.									0
				Cumulative Total					90

NOTE : The concept of 'becoming' has been coded as : Always in process plus self-actualizing.

38

**TABLE 2 BONNER'S VIEW OF MAN:
PERCENTAGE EMPHASIS**

No.	Dimension/Category	Total	Percentage Degree of Emphasis
1.	A communicative being	0	0
2.	Accepting of non-hedonic emotions	2	2.2
3.	Always in process	7	7.8
4.	Conscious/Aware	2	2.2
5.	Creative	2	2.2
6.	Forward-thrusting	5	5.6
7.	Freedom-cherishing	1	1.1
8.	Future-imagining	2	2.2
9.	Goal-directed	4	4.4
10.	Holistic/An integrated whole	4	4.4
11.	Intelligent	0	0
12.	Much like fellow man	0	0
13.	Nature-appreciating	0	0
14.	Ontologically responsible	3	3.3
15.	Open to life/experience	2	2.2
16.	Other-affirming	4	4.4
17.	Pleasure-loving	0	0
18.	Present-confronting	0	0
19.	Rational	0	0
20.	Risk-taking/Courageous	2	2.2
21.	Self-active	0	0
22.	Self-actualizing/Self-realizing	14	15.6
23.	Self-affirming	4	4.4
24.	Self-determining	17	18.9
25.	Self-disciplining	3	3.3
26.	Sexual	0	0
27.	Socially responsible	5	5.6
28.	Ultimately unknowable	1	1.1
29.	Unique	6	6.7
30a.	Reality-accepting	0	0
30b.	Ultimately alone	0	0
		90	99.8*

* This amount differs from 100 percent because the figures were rounded off to the first decimal place for the sake of clarity.

39

SUMMARY

This chapter has consisted of the summarized version of Bonner's view of man and its analysis by means of the specially designed content analysis instrument.

In his view of man, Bonner has given primary emphasis (10 percent or more emphasis) to the following dimensions :

Man is self-determining

Man is self-actualizing

The areas of secondary emphasis (between 5 and 10 percent emphasis) were :

Man is always in process

Man is unique

Man is forward-thrusting

Man is socially responsible

The remaining dimensions received tertiary emphasis (less than 5 percent emphasis).

It is interesting to note that the concept of man being forward-thrusting or proactive received only secondary emphasis since the sub-title of Bonner's book *On Being Mindful of Man* is Essay Toward A Proactive Psychology.

One other discrepancy needs to be highlighted. Whereas Bonner explicitly states, on the first page of this summarized version of his view of man, that man is a blend of the conceptions of responsibility and the full actualization of human potentialities, the content analysis of his view reveals that he gives only secondary emphasis to man being socially responsible. This indicates that, despite his stress on social responsibility and other-mindedness Bonner sees these features as an epiphenomena of man's primary self-actualizing, ceaselessly self-perfecting nature.

5

JAMES F.T. BUGENTAL'S VIEW OF MAN

Bugental's view of man has been obtained from his book, *The Search for Authenticity*. The four parts of the book have all been carefully analyzed and relevant passages from them are quoted below to form a summarized version of his view.

...We do not know very much yet about what it means to be a human being. We do not appreciate in any depth what the potentials of human imagination, creativity, and variation may be. We are more governed by superstition than fact, more limited by tradition than recognition, more inhibited by our own fears than by external constraints in exploring the whole world of our being. I believe we have made but the barest beginning on exploring this new hemisphere, this human frontier...[1]

In the more than two decades since World War II began, educated thinking in America about man and his condition has undergone tremendous changes. Twenty-five years ago, psychology, psychiatry, psychotherapy, and related disciplines were barely emerging into a wholistic conception of the human being. By and large the dominant influences in these fields were reductionistic, mechanistic, and part-function centered. In the intervening time, these influences seem to me to have reached and passed their floodtides. The orientation is, to be sure, still very much employed and productive, but a newer, more inclusive perspective on the human experience is growing rapidly and appears to be the ascendant one. This emerging

---

[1] James F.T. Bugental, *The Search for Authenticity* (New York : Holt, Rinehart and Winston, Inc., 1965), p. 9.

orientation has been called variously, "the third force" in psychology (after psychoanalysis and behaviorism) "neo-phenomenology," and "humanistic psychology."[2]

The following pages will make a beginning of an affirmative statement of the nature of the humanistic orientation in psychology. I will do this by setting forth five postulates of humanistic psychology that seem to represent common elements in the perspectives of most writers identifying with this field... We are only now beginning to discover the commonalities in the diverse spokesmen of the humanistic perspective. It is probable and highly desirable that the list of postulates that follows will be criticised, revised, and supplemented many times.

Man, as Man, Supercedes the Sum of his Parts

When we speak of "man" in humanistic psychology, we do so with the intent of characterizing a person rather than an "organism." Humanistic psychology is concerned with man at his most human or, to say it differently, with that which most distinguishes man as a unique species.

Our first postulate states the keystone position that man must be recognized as something other than an additive product of various part-functions: Part-function knowledge is important scientific knowledge, but it is not knowledge of man as man. It is knowledge of the functioning of parts of an organism.

Man has his Being in a Human Context

We postulate second that the unique nature of man is expressed through his always being in relationship with his fellows. Humanistic psychology is always concerned with man in his interpersonal potential. This is not to say that humanistic psychology may not deal with such issues as man's aloneness, but it will be evident that even in so designating it, "aloneness," we are speaking of man in his human context. The psychology of part-functions is a psychology that mechanically and incompletely handles this relatedness (actual or potential) of the human experience.

---

[2] Bugental, *op. cit.*, pp. 9-10.

Man is Aware

A central fact of human experience is that man is aware. Awareness is postulated to be continuous and at many levels. By so viewing it, we recognize that all aspects of his experience are not equally available to man but that, whatever the degree of consciousness, awareness is an essential part of man's being. The continuous nature of awareness is deemed essential to an understanding of human experience. Man does not move from discrete episode to discrete episode, a fact overlooked by experiments of the behavioristic orientation when they treat their subjects as though they had no prior awareness before coming into the experimental situation. Our postulation also provides for unconsciousness as a level of awareness of which there is not direct apprehension but in which awareness is nevertheless present...

Man has Choice

There is no desire here to resume the hoary debate regarding free will versus determinism. Phenomenologically, choice is a given of experience. As man is aware, he is aware that his choices make a difference in the flow of his awareness, that he is not a bystander but a participant in experience. From this fact flows man's potential to transcend his creatureliness which is also to say that from this postulation we derive man's capability of change.

Man is Intentional

In his choices, man demonstrates his intent. This does not mean "striving," but it does mean orientation. Man intends through having purpose, through valuing, and through creating and recognizing meaning. Man's intentionality is the basis on which he builds his identity, and it distinguishes him from other species.

The characteristics of man's intentionality need to be specified: man intends both conservation and change. Mechanistic views of man frequently deal only with drive-reduction and homeostatic conceptions. Humanistic psychology recognizes that man seeks rest but concurrently seeks variety and disequilibrium. Thus we may say that man intends multiply, complexly, and even paradoxically.[3]

We are ready now to examine the central thesis that will be presented in this book. As this is an existential point of view in a humanistic setting, our concern will be to find those ways of conceptualizing the human experience that will facilitate our efforts to release and facilitate human potential. As of now, this is how I see this matter.

I will postulate that the primary value in human life is to live in accord with (indeed, as a part of) the way things really are. Of course, that is a hazardous thing to say, for who can claim to know how things really are? The answer is, nobody and everybody. Nobody can prove he knows for sure, and everybody acts on the assumption that he knows. Probably, almost surely there is no one "way things really are," there are only various ways of seeing our situation...

Now, just as I have postulated a primary value, I will point to a primary human process: awareness. Through awareness we discover ourselves and our world. Through awareness we can estimate our relation to world. It seems to me useful to think of that relationship as having these four characteristics:

We are limited in our awareness of ourselves and of world.

We can act in ways that affect our awareness of ourselves and of world.

We have choice about which actions to take and not to take.

While each of us is in one sense alone, in another we are all related.

Throughout the whole of this thesis the fundamental concern is with authenticity of being. The distortions of being that give rise to the need for psychotherapy are inauthenticities. The celebrations of being that ontogogic therapy seeks to facilitate are the products of authenticity. The influence that makes the change possible is the response of awareness to the authentic in life.[4]

---

[3] Bugental, op. cit., pp. 11-13,

[4] *Ibid.*, pp. 14-15.

....

A person is authentic in that degree to which his being in the world is unqualifiedly in accord with the givenness of his own nature and of the world. Authenticity is the primary good or value of the existential viewpoint.

The value of authenticity is postulated rather than argued...

....

When our being-in-the-world is in accord with the nature of ourselves in the world we are authentic. *Authenticity* is the term 1 will use to characterize both an hypothesized ultimate state of at-oneness with the cosmos and the immense continuum leading toward that ultimate ideal.[5]

....

Authenticity is a term used to characterize a way of being in the world in which one's being is in harmony with the being of the world itself. To say it differently, we are authentic to that degree to which we are at one with the whole of being (world); we are inauthentic to the extent that we are in conflict with the givenness of being. Clearly, I am here seeking to characterize an ideal or ultimate condition of authenticity with the recognition that we are always somewhat less than fully authentic. Also implicit in my description of authenticity, and of much more conceptual significance, is the recognition that authenticity amounts to the resolution of the subject-object split, the self-world dichotomy. As one approaches the stage of letting go to the suchness of Being without striving against it, one is attaining to full authenticity.[6]

....

Fundamental to all else in the human experience is awareness. At least phenomenologically, world arises out of human awareness. It grows and evolves with experience. Learning importantly affects what one may become aware of and what will escape awareness. Emotional needs have similar effects.[7]

....

Awareness discloses to each person (*a*) that he is finite, (*b*) that he has the potential to take action, (*c*) that he has some

---

[5] Bugental, *op. cit.*, pp. 31-32.

[6] *Ibid.*, p. 33.

[7] *Ibid.*

choice of what action he will take, and (*d*) that he is at once separate from, yet related to, his fellows. From these flow, respectively, the existential anxieties (*a*) of fate and death, (*b*) of guilt and condemnation, (*c*) of meaninglessness and emptiness, and (*d*) of loneliness and isolation.[8]

....

Our study of the nature of awareness has provided us with four aspects that may be identified for heuristic purposes. These are spelled out solely to aid our understanding and not in the belief that each is independent of the others or that this is necessarily the one correct way of analyzing awareness. Awareness is a phenomenological whole and may be analyzed into various components according to the needs of the situation...

....

We have now a first basis for understanding the nature of authentic awareness. It is awareness that fully confronts the existential anxieties of being and affirms its own being by incorporating those anxieties while yet avoiding their distortion. Man lives in contingency. He can and does take action that affects his awareness and experience. He takes such action without ultimate guide posts of universal values or built-in instincts. And man is in constant relation with his fellows while yet being separate from them.

Of course, man does not ever fully accept the givens and their attendant anxieties. Instead he feels compelled at times to deny some parts of his condition and to try to nullify their effects. When he does so, he distorts his situation and becomes prey to neurotic anxiety and inauthenticity...[9]

....

What are the attributes of being authentic?
1. Being as fully aware as I can be at the moment.
2. Choosing what possibility I will invest with my life, with actuality, at the moment.
3. Taking responsibility for the choice I have made while yet recognizing the imperfection of my awareness and the fact that my choice gave this alternative actuality and not some other.

---

[8] Bugental, *op. cit.*, p. 30.
[9] *Ibid.*, pp. 39-40.

Recognizing therein that tragedy is always potential and that neither my limitations of awareness, nor my good intentions, nor my suffering, nor my virtue, nor any other extrinsic circumstance, can change that fact.[10]

Quotations 11 thru 15 are from part two of the book.

What is evident when we step back from our preconceptions of habit and language is that the business of being a person, a human being, is enormously complex and endlessly varied. Most of us were taught much more about how to manipulate certain relatively infrequently used mathematical functions than about how to make certain crucial life decisions.[11]

. . . .

Man finds himself in "a world he never made," a world of apparently infinite possibility. In this world, man is constantly confronted with choices for which he must always be less than adequately prepared. Thus man is constantly faced with uncertainty, with contingency, with the possibility that his choices may fail to bring him the results he intends. That failure may range from the simplest disappointment to a fatal misstep, from being late to a pleasant evening's recreation to being killed as he walks across his normally quiet, residential street.

This contingent plight is the constantly present and always overriding fact of our existence ...[12]

. . . .

There seems to be a kind of mythology of the naturally good man involved here, a mythology that is not truly humanistic. I prefer to envisage the situation as one in which there is no instinct toward either good or evil. The concept of instinct robs man of choice. I am convinced that when we are free of distortions we are able to see existence in such a way that we will make choices which are in harmony with life in general, because these become the choices that are most fulfilling to us as individuals. This is akin to the concept of enlightened self-interest but is not synonymous with that concept.[13]

---

[10] Bugental, *op. cit.*, p. 45.

[11] *Ibid.*, p. 55.

[12] *Ibid.*, p. 74.

[13] *Ibid.*, p. 83.

....

Every therapist needs to examine his own conscious and explicit as well as his implicit beliefs about this matter of the basic nature of man. Many times, if he is candid with himself, the therapist will find that — at least for himself and his family — he acts in terms of a traditional and nonhumanistic view of man. Freudian psychology and much of Christian psychology have taught the view that man's impulses are bad or dangerous and have to be restrained and redirected. Striving and inhibition are key concepts — even values — in such a perspective. Control is not for artistic modulation but for protective restraint.

Some psychotherapists insist the reverse: man is basically good; only release him in every way possible and all will be well. To me, these seem the familiar mistrust ideas inverted. Such assertions have the ring of protest and, some times, of overdetermined blindness to the world of actuality.

It seems to me that existentially we must recognize that "good" and "bad" are not characteristics of the given world of being. Were they so, we would have less choice than I believe is our condition. If one can genuinely counsel that man is inherently good, then all we need is an improved technology for releasing man, and our real freedom is at an end. The best technology becomes the operational equivalent of the ultimate good, and man is but an object of that technology and his inborn goodness. No, I do not believe it.

Man is. Not man is good or man is bad. Within the limits of his awareness, responsible for his choices, without the guideposts of instincts or innate nature, and trying to find meaningful relation with his fellows, man is, period.[14]

....

Man finds himself thrown into a world of infinite possibilities where each moment is a choice point, each act gives life (actuality) to some possibilities and condemns others to oblivion. From each such choice branch unimaginable consequences. It is this very unimaginable quality that is at the base of man's living in contingency. Contingency here

---

[14] Bugental, *op. cit.*, pp. 142-143.

means that any act has an infinite array of possible outcomes and man can at any point recognize only some finite part of this infinite array. Man organizes his choices, his adaptations, in terms of his estimation of the possibility of various outcomes and tries in the process to increase his chances of actualizing that which he wants in his life, while diminishing the probabilities of that which he does not want. The impossibility of assuring the outcomes of one's actions, the consequences of one's choices, means that always there exists the possibility of tragedy ensuing from any choice which one makes. This is a possibility against which we can never be completely insured. Tragedy, then resides in man's nature. Tragedy resides in the fact that we can never know enough to protect ourselves and those whose welfare we cherish against the unfortunate consequences of our own conscious and unconscious choices and actions.[15]

Quotations 16 thru 44 are from part three of the book.

Repeatedly in this book I will insist that recognition needs to be given to the manifest reality that feelingful awareness is the basic process of being...[16]

. . . .

The essential affirmative assertion to be made here is that awareness is always feelingful awareness and only neurotic distortion (resistance) or the artificial models of awareness developed for research or teaching (God help us!) divide the cognitive from the conative and the affectional. Awareness, we are saying, is an integral process, which - for heuristic purposes - we may describe as having aspects of meaning, feeling, and action potential, but awareness always has these aspects and is not simply one or the other.[17]

. . . .

Awareness is an evolving phenomenon. . .

Awareness grows. It is not a static, given quantum...it is an

---

[15] Bugental, *op. cit.*, pp. 152-153.

[16] *Ibid.*, p. 217.

[17] *Ibid.*, p. 220.

evolutionary sequence — for the race and for the individual.[18]

....

Ontological freedom is the name I suggest for that realm of greater realization of our potential which is latent to every one of us ...

When we come to conceive what is meant by genuine ontologic freedom, we find that conceptions are possible at a number of different levels, ranging from the simple opening up of our familiar perception of being to quite metaphysical and speculative reaches. In this chapter and the next I will present three conceptualizations of the nature of ontologic freedom. In sequence these will be termed: emancipation, actualization, and transcendence. These titles are used solely for our ease in reference. They describe ways of grouping our observations (not discoveries) of systems of ontologic freedom. The conceptualization of ontologic freedom is capable of an infinite variety of approaches. The three here employed seem useful in aiding communication, but they are by no means advanced as the only ways of grappling with this difficult problem.

It will be helpful briefly to characterize each of these three forms of conceptualization before going further. *Emancipation* will be used to refer to the process of freeing one's awareness of being from the false equation of the *I* and the *Me*. *Actualization* carries the understanding of freedom into a different area as it suggests the greater scope of choice available to the person in expressing his being in the familiar world of experience. "Actualization" here is used in the same general sense as it is used by Maslow or as "fully functioning" is used by Rogers. *Transcendence* brings us to the realm of metaphysical speculation; probably the nearest equivalent of transcendence as here conceived would be the Buddhist concept of *Satori*.[19]

When a person is able to free himself of the false equation of the *I* and the *Me*, he is open to experience his life in a new fashion... There are a number of concrete ways in which this emancipation is experienced and from which greater choice can flow. In essence what emancipation means is that

---

[18] Bugental, *op. cit.*, pp. 220-221.
[19] *Ibid.*, pp. 253-254.

the patient can recognize himself now as the subject of his experience instead of the object.[20]

. . . .

What would it mean to be freed of the *Self*? It would mean living fully in each moment, dwelling on the "razor's edge" of Now. This is such a basically unfamiliar concept that it is easily dismissed too lightly. . .

To be freed of the *Self* would be to meet our lives with "new innocence," to choose in terms of the intrinsic properties of the alternatives, to enter into dialogue with another person without pretense and with genuine contact with him. It would mean a keener edge to experience, an intensified involvement, more acute feelings whether happy or sad (but not more mood extremes), more detachment from the trivial and popular, more risking in all of life.[21]

. . . .

Some time ago the *New Yorker* ran a cartoon of a puppet that manipulated its own strings. This was not, to my mind, a funny cartoon. It graphically portrayed a disease endemic in our American middle-class culture: alienation from one's own being. The alienated person breaks himself up into a tragically ludicrous self-manipulating puppet. He has lost the immediacy of being that is the birthright of human life.

Let us examine the sequence which brings about such alienation in order that we can understand its meaning and contrast it with the emancipation of ontological freedom.

Experiencing anxiety, existential or neurotic, one may find the discomfort such that he attempts to "shut it off." I have said that the awareness which is our basic process of being is *feelingful* awareness. Anxiety is experienced through the feelings, and so the "shutting off" is a matter of shutting off the feelingfulness of our awareness...

...Cut off from direct experience of his feelings, such a person comes to treat himself as an object...The "shutting off" of the feelings cannot be done for just the anxiety, just the unpleasant, unwanted feelings. Feelingfulness is really all of a piece; the savor as well as the threat of living is suppressed.

---

[20] Bugental, *op. cit.*, p. 254.
[21] *Ibid.*, p. 323.

The alienated person typically complains that he does not have much joy, can not be sure that he loves those he is close to, wonders if he really likes what he does.

The sorry sequence has one further extension. The sense of relatedness with others depends in appreciable part upon the feelingfulness of awareness. Experiencing another with empathy gives the person an appreciation of the deep bond which bridges our aloneness. But the alienated person, treating himself as an object, soon is treating those about him so also. This is Buber's "It-it" relationship. Like billiard balls clicking against each other, not like human beings interpenetrating with each other, such relationships are frustrating and incomplete.[22]

Actualization is the name I am suggesting for a way of being in one's life in which there is greater realization upon the potentials of human existence than is usual, at least in our culture...[23]

Almost without exception, the person who moves toward greater realization of his own potential shows a succession of changes in his feelings of concern. Both the substance of what elicits concern and the form in which that concern is expressed show this evolution. . .

. . .In a somewhat oversimplified fashion, we may say that the person attaining to actualization is concerned significantly with much in his life but that no concern seems crucial.

Authentic concern seems to be a matter of being willing to make commitments, to let things matter, and yet to retain sufficient perspective so that one is not catastrophically imperiled or even overcome by reverses in any one or any group of concerns. This characterization picks out another aspect of neurotic concern which may be contrasted with the authentic: the neurotic often is so threatened by his restricted, covert concern that he is unable to allow himself investment in that which is truly important to him.

. . .The changed concern of authentic being is that concern is directed toward that which one is intending, not dispersed

----

[22] Bugental, *op. cit.,* pp. 260-261.

[23] *Ibid.,* p. 263.

on what one may be disclosing or other irrelevant matters. What we have just pointed out about authentic concern being directed toward what one is intending instead of being dispersed on extraneous matters is an example also of a second frequent characteristic of the person coming to greater actualization of his being: the ability to commit himself selectively and effectively. The authentic person seems to have at his conscious direction the energies and emotional involvement that may be scattered and uncontrolled in the neurotic.[24]

What I have been saying about the changed nature of concern and the selective commitment which characterize the person achieving greater actualization of his being will make apparent why we will usually find that such a person participates with much more whole-hearted involvement in those matters to which he does choose to devote himself.[25]

Authentic living is characterized by the dropping away of happiness as a goal in itself. The actualizing person is busy with the concerns to which he has chosen to commit his living and seldom stops to assess his happiness. Very often, of course, it will be apparent that he is a happy person - although by no means is this always so - but it seems to be only the neurotic and the unhappy that expend their concern explicitly and directly on their happiness. The impression is inescapable that happiness is a state that is pushed away by the hand that would grasp it but that tends to accompany the person who is alive to his own being.[26]

The authentic person seems to have a perspective on life which enables him to take satisfaction in the very fabric of being...

The person who is actualizing his own being seems to have a style to his living, a certain artistic quality. Often this quality is not externally obvious, but to intimates it is apparent...[27]

---

[21] Dugental, *op. cit.*, pp. 266-268

[25] *Ibid.*, p. 270.

[26] *Ibid.*, p. 272.

[27] *Ibid.*, pp. 272-273.

. . . .

. . .the wholeness of experience is discovered on a broader scale: a recognition of the essential humanness of those who formerly seemed so different or so hostile, an empathy with the human experience of people in general, a feeling for one's own participation in the general and of the universal human outlook being present in oneself.[28]

. . . .

Except for its immense importance, this over-coming of the subject-object split might be subsumed under the recognition of the wholeness of experience. It is certainly a part of that recognition. Discovery of one's essential rootedness in all mankind gives a sense of being uniquely oneself while yet related to all others. World no longer seems out there and in opposition; rather the boundary between *I* and world is experienced as fluid and changing with one's awareness and experience... Similarly, one no longer feels the split within himself. . .[29]

Centeredness

This is Rollo May's term to express an awareness of being in one's own life. It is expressive of the direct intuition of being which replaces self-consciousness when that handicapping sort of awareness is surmounted. The centered person (not *Self*-centered) is the person who is actualizing his being with aware choice. He recognizes the possibilities open to him, the contingencies that are within his ken, the responsibility of commitment, and he chooses knowingly.

Centeredness is the positive aspect of the same process I have characterized as the freeing of the *I* from the false equation with the *Me*. As this handicapping linkage is dissolved, the person becomes centered in his being in the moment of actuality and leaves behind the diffuse concern with the past or fruitless apprehensions of the future that characterize the inauthentic. Actualization occurs only in the moment, Now. Centeredness is the active core of actualization.[30]

---

[28] Bugental, *op. cit.*, p. 274.
[29] *Ibid.*, p. 275,
[30] Ibid.

Transcendence is an hypothesized point of full authenticity of being in which the person would emerge into oneness with The All. There is no nonmystical language really available for describing this concept. By its very nature it partakes of mystery or mysticism - it is unknown to our usual understanding. Transcendence implies the complete confronting and incorporating of existential anxiety in all its forms. It includes, by definition, the overcoming of the subject-object split both within the individual and between the person and his world. Transcendence is complete awareness and full feelingful assent. We choose to include the concept of transcendence in our thinking because it is essential to put into perspective the other forms of ontologic freedom, emancipation and actualization. When these are set against the background of transcendence they may be seen as the significant but incomplete forms of being that they are. This is by no means to devalue the meaningfulness of emancipation and actualization. Each represents a truer affirmation of life than most of us now know. But transcendence provides the further value of opening our awareness to the true immensity of our potential...[31]

The basic fact of existence is existence. The basic significance of existence is the potentiality of being. The basic experience of existence is awareness. This much is the framework within which I write; all that I say hereafter is derived from the intercourse of awareness with existence, is the creation of our being from the is-ness of existence. Existence is, period. One can say no more of existence, as existence. The meanings and descriptions we write are neither purely discoveries about the givenness of existence nor yet are they purely our inventions imposed on existence. They are the outcomes of our being and of our being aware as a part of existence.[32]

When man encounters the givens of his existence he experiences existential anxiety. How man responds to that anxiety - whether with dread or with courage - tells the story of his

---

[31] Bugental, op. cit., p. 277.
[32] *Ibid.*, pp. 282-283.

non-being or his being. The crossroads of life, existentially speaking, lie at the point of the confrontation of existential anxiety...[33]

. . .It [the core dynamic sequence through which the person comes to experience actualization or blockage of his existential needs] may be read schematically as follows : The existential givens of our being, once recognized, occasion deep feelings of existential anxiety. That anxiety is natural to our being, but it may seem too overwhelming at times. Genuine confrontation of existential anxiety means taking into our very awareness of ourselves certain attributes of being in the world that may seem more than we can sustain. In such instances we may try to avoid being overwhelmed by distorting the nature of the givens of existence. When we do so, we experience dread and the feelings of neurotic anxiety. If, on the other hand, we confront existential anxiety authentically and take into our experience of ourselves the aspects of being that seem so threatening, we are making the courageous response and are freed for authentic being. Only when we are authentic in our being can we truly satisfy our basic existential needs. . .[34]

....

THE CORE DYNAMIC SEQUENCE[35]

I discover world through awareness. I am in world.
I am. . .
FINITE . . . ABLE TO ACT . . . ABLE TO CHOOSE. . . SEPARATE
These are the *Existential givens.*
Because I am so, I find I am subject to. . .FATE. . .GUILT. . .EMPTINESS. . .LONELINESS
These are the forms of *Existential anxiety.*
I cannot escape existential anxiety. I can confront it.
To confront it means to incorporate into my being-in-the-world. . .

---

[33] Bugental, *op. cit.,* p. 286.
[34] *Ibid.,* pp. 286-288.
[35] *Ibid.,* p. 287.

CONTINGENCE. . .RESPONSIBILITY. . . AUTONOMY . . .APARTNESS

These are the *Existential confrontations.*

If I find these too devastating to accept, I may seek to avoid existential anxiety. Thus I will fall prey to feelings of ...POWERLESSNESS...BLAME...ABSURDITY...
ESTRANGEMENT

These are the forms of *Neurotic anxiety or Dread.*

On the other hand, if I do confront and incorporate existential anxiety, I am able to realize my being in the world through...FAITH...COMMITMENT...CREATIVITY...LOVE

These are the forms of *Authentic being or Courage.*

If I am authentic in my being in the world, then I am able to realize. . .

ROOTEDNESS...IDENTITY...MEANINGFULNESS...
RELATEDNESS

These are the *Existential needs.*

....

The four forms of existential anxiety that have just been detailed are the cornerstones of the conception here being set forth. Each person must deal with each of these forms of anxiety in some fashion. Obviously, the manner in which one handles his experience of such anxiety will vary tremendously from person to person. However, as we have seen, there are two general ways of responding to the experience of existential anxiety. These are the response of dread and the response of courage.

The response of *dread* is that which gives rise to neurotic anxiety. When the recognition of existential anxiety seems overwhelming and the person cannot support incorporation of that anxiety, then he must engage in maneuvers designed to distort the reality of his awareness of being. His experience of existential anxiety is transmuted, as it were, into dread, which in turn produces various forms of neurotic anxiety.

When a person confronts existential anxiety and accepts it and incorporates it within his being, we speak of his response being that of *courage*. In such instances the forms his response takes are fulfilling of his existential needs. Saying it differently, we are postulating that *to recognize that one lives in anxiety is a part of the response to that anxiety which makes it possible for one then to recognize his basic needs as a human being and set about fulfilling them.*[36]

. . . .

Poised in the apparent chaos of raw contingency, finding no solid footing on any side, discovering all "realities" to be only probabilities, if I am aware and am aware only of contingency, then - quite literally - I will die or go mad. It does not seem likely that such a confrontation is endurable. . .

. . .the anxiety attendant upon contingency, the existential anxiety of fate and death, may be incorporated and a rootedness achieved that makes for greater authenticity in being. How is this possible? The answer is what I have chosen to term "the response of faith."

Faith, as I use it here, is intrinsic faith. It is objectless faith. It is the *I* affirming its own being. It is a confronting of the infinitude of contingency with the declaration, "I am I. This is my starting place. This is my certainty, though there be no other."[37]

. . . .

I find it useful to think about the process of living as that of a creative, artistic enterprise, not unlike the painting of a picture. If we use this analogy in the present instance, we will see readily that the artist who hesitated to put an end to his picture, who created a masterpiece but then added canvas to canvas again and again and kept painting, would in time destroy his picture. Saying it differently, a part of artistry is knowing when to stop. Part of artistry is knowing what to include and what not to include. Part of artistry is knowing where the frame goes around the picture, where the edge of the canvas is. Those who engage in photography will know how important it is in composing adequate pictures to frame them properly, to know the limits, to set the limits by choice and not by chance. So it is with life; there are many opportunities for limits that can give artistry, grace, dignity, meaning to our living. If we are authentically in our lives, we use the opportunity to set limits esthetically and vitally. Of course many of us—and I certainly include myself among these - are not so creative. We are fearful of limits because we feel if we do not take all we can

---

[36] Bugental, *op. cit.*, pp. 289-290.

[37] *Ibid.*, pp. 328-329.

now, may never get any more. In this fashion we can spoil so much that could be rich in our lives. . .[38]

The exercise of my potential to take action affects what will be present in the stream of my awareness. What is in the stream of my awareness makes an emotional difference to me. The experiencing of this emotional difference we call responsibility. Responsibility is the subjective correlate of the existential given of the ability to take action. Thus it leads to the existential anxiety of guilt and condemnation. I am not indifferent but concerned about my actions and their consequences. I recognize that my doing and not doing are importantly involved and that I must take the responsibility for such doing and not doing. Guilt and condemnation certainly evoke anxiety, but they are the expression too of the fact that I live my life rather than being lived by it. In my acceptance of the anxiety of guilt and condemnation I affirm my identity.[39]

Commitment is, in paraphrase, the statement, "This I am; this I believe; this I do. I am the being, the believing, the doing." Commitment is not the place in which one stands, the tenet one believes, or the act one does, however. Commitment is not a subscription to something external to the person's own life, no matter how worthwhile. Commitment, as we are using the term here, is not to world peace, nor to the prevention of juvenile delinquency, nor to mental health, nor even to one's own future as such. Commitment is an awareness, an attitude, a clear and feelingful recognition of being fully present in the moment, making the choices of the moment, and standing by the consequences of those choices whether anticipated or not. Commitment is "playing for keeps" rather than vainly pleading for "slips (to) go over," as do small children in their games of marbles.

Authentic commitment is possible to the person who genuinely accepts responsibility in his life...[40]

---

[38] Bugental, *op. cit.*, pp. 332-333.
[39] *Ibid.*, p. 334.
[40] *Ibid.*, p. 338.

. . . .

. . .The experience of being created, being a creature, by itself would lead to the anxiety of absurdity and meaninglessness were it not that the experience of choice opens the possibility of transcending this creaturely status. As man exercises choice he takes part in creation and overcomes his object status to become subject in his world of experience. As subject and creator man creates meaningfulness. This is a supreme achievement of man's choices: the creation of meaning where there was the threat of emptiness and the potential of absurdity.[41]

. . . .

. . .World, we discover, has few constraints. Perhaps only the four we have called the existential givens plus the physical limitations of gravity and hunger and such. World is so open. For some that openness is the terrorizing lack of meaning and control which makes it a rainy free-way without lines. For others it is the openness of a fresh canvas awaiting the artist's brush, a keyboard potent with music, a vast mountain range to be explored. The difference seems to be that the latter group responds to the emptiness with creativity, the former see only absurdity.

Creativity means not simply the public creations of artists and artisans. It means, more importantly, the inner creativity that is potential to each person. If refers not to the product created but to the act of creation. To the extent that one makes his choices out of his own being and with faith in being, to that extent is he creative whether or not that which he produces has been made a million times over. To the extent one patterns his choices on that which is external to his own being, one is not truly creative although the product is hailed as unique by all who see it.[42]

. . . .

Man seeks relationship with man as a way of dealing with the fact of apartness and giving expression to his being a part of life. Separateness in itself has no emotional tone, but when experienced only in its lonely aspects it leads to the experience

---

[41] Bugental, *op. cit.*, p. 344.
[42] *Ibid.*, pp. 347-348.

of loneliness. Relatedness is the concurrent need of man to give an answer to his condition of separateness.[43]

"Love, the answer to the problem of human existence," is the title of a section in Erich Fromm's *The Art of Loving*. While I might take issue with calling human existence a "problem," I do not debate the central significance of love. . .

Existential love is an expression of one's whole being in relation to all Being. Existential love is — in its most transcendent form — participation in all Being, participation so complete that the subject-object dichotomy is obliterated and the essential unity of The All is revealed. Clearly, in such transcendent love, faith, commitment, and creativity are ultimately expressed. In speaking of transcendent love we are characterizing an ultimate realization of the human potential, a point at which apartness is absorbed into wholeness. Most of our concern, however, is with love as a response to the experience of apartness, as a confronting and incorporating of the existential anxiety of loneliness and isolation. This brings our attention to a realm of more familiar experience. Here we may speak of actualizing love. Actualizing love is the "I-Thou" relation which Buber has characterized so well. It is the affirmation of one's own being in relation to another. Perhaps, in the terms we have been employing, it is the relation of *I-process* to *I-process*. . .[44]

Quotations 45 thru 52 are from part four of the book.

Change is coming, change is occurring, change in the very center of the human experience, change in the very conception of what it means *to be a man, to be alive, to be at all*. Already the effects of this tremendous enterprise are being importantly felt in religion, in education, in the arts, in child-rearing, in literature. There are modest evidences of the wave reaching to politics, commerce and industry, the military, international relations. And the wave that is now spreading is really only

---

[43] Bugental, *op. cit.*, p. 352.
[44] *Ibid.*, p. 353.

a forerunner from the earlier stages of the investigations. The waves that will follow will be so much greater that the presently visible changes, pervasive as they are, will be seen to be but ripples. In the simplest terms, I think our whole conception of human life is in the process of being completely revised. Since from that conception ultimately flows all else that we know, I think the whole experience of life is under revision as well.[45]

I want to describe four observations about the attributes of being psychologically healthy, of truly realizing one's being, of emerging as a man. For ease in reference, I will call these: presence, valuing the full emotional range, artistry in living, and dynamic awareness of being.[46]

Presence

I have already described the significance existentialists assign to this term, but let me restate it to make evident the particular coloring which I here want to emphasize. Presence is the quality of immediate awareness in which one knows directly his own being in relation to his situation. By "knows directly" we mean to indicate an implicit awareness, an intuition, as contrasted with a reflexive observation. In terms we have developed earlier, we are speaking of *I-process* awareness as opposed to *Self* awareness. Presence is "being there" (dasein) in the purest sense. . .[47]

The emergent person is jealous of his own presence in his own life. That is to say, he seeks to know as fully as possible what it is he is doing, what is his role in bringing about the experiences he has and what the feelings are that he undergoes. . .[48]

Emotional Fullness

. . .The idea of valuing the entire range of our emotions from

---

[45] Bugental, *op. cit.*, p. 377.
[46] *Ibid.*, p. 382.
[47] *Ibid.*, p. 383.
[48] *Ibid.*, p. 386.

the most unpleasant to the most satisfying may seem strange to some. And yet I am convinced that the more a person is truly actualizing his being, then the more he is open to and appreciative of the entire range of his emotions.

A simple analogy may illustrate this point: If I am piloting a plane and certain gauges give me unwelcome information, such as that the wing is icing or that there is undue heat in a certain motor, I would be foolish indeed to disconnect those meters which inform me of these facts. It would be ridiculous to argue, "Well, what they tell me makes me uncomfortable; therefore I'll cut them off." I am sure the parallel that I have in mind is evident. When we suppress awareness of pain or fear or any other emotion, we run a similar risk of depriving ourselves of information essential to our well being.

Every once in a while one reads of a child born without a sense of pain. It is no surprise to learn that such an infant has a very poor life expectancy. Far from being freed from a curse, he is condemned and cursed indeed in having no proper warning system to use in self-protection. Those of us who are successful in suppressing an emotion, say our sense of fear, place ourselves in similar jeopardy. . .

There is a further advantage to cherishing the entire emotional range. In one sense we can think that ultimately each of us is most motivated, not by reason, but by feelings, that what we seek is feeling a state of pleasure, of comfort, of realization or achievement, or of quiet presence. Whatever the names given to the feelings, these seem in their way more ultimate than all the reasons, all the logic that may serve us on our way to the feeling state. This is true even of the martyr who may undergo a hell of feelings in order to maintain what for him is even more important, the feeling of being faithful to himself and to that in which he believes. When we have clear awareness of all of our own feelings readily available, we are enabled to make sounder choices and follow through on those choices more dependably.

Still another healthful gain in cherishing the range of our feelings is that in this fashion our reality awareness is heightened. It can safely be said that much of the distortion of reality characteristic of neurosis is brought about by the attempt to deny or alter certain feelings the person fears to have. When he seeks to avoid the painful emotion, he often finds himself distorting what is real about himself and his world. In this way new confusion and new needs to distort are produced. . .

. . .One distortion leads to another, and increasingly the sense of presence in one's own life and the feeling of vitality in one's own experience get lost.

The emergent man values a wide range of emotional potential that provides him with significant information by which to orient himself, that makes possible life decisions on which he can follow through responsibly, and that helps him to keep his awareness open and undistorted. Needless to say, too, such an emotional range gives life more color and meaning. . .[49]

Artistry in Living

. . .People who have had a successful psychotherapeutic experience or those people who without this experience demonstrate richness of personal living show a kind of artistry, a type of beauty in the manner in which they arrange their lives. Typically we are impressed with such a person's ability to select, to act from choice rather than from compulsion; typically there is a sense of the harmony among the various parts of such a person's living.

. . . .

. . .it is reported that the creative person has the capacity to work at a problem with great intensity until he hits a snag. Then he "turns it over to his unconscious" while he undertakes something else. Later when he picks up the blocked task again, he often finds that he has new insights into its solution which are clearly the product of unconscious work during the recess from conscious effort. I am confident that the creative person, the emergent personality, is one who makes rich and frequent use of the resources of both conscious and unconscious life.[50]

Dynamic Awareness of Being

. . .living, at its most real, is a creative process. It is an artistic process in which in some measure we choose our way of being at each moment. Psychotherapy is not a matter of

discovering static, predetermined facts about oneself. Rather,

---

[49] Bugental, *op. cit.*, pp. 386-388.
[50] *Ibid.*, p. 390.

truly useful, intensive psychotherapy has almost the contrary aim: it seeks to release the person from the static *Self* to realize his true freedom.

This is a difficult understanding to come by. You and I have learned to think of ourselves and other people as being thus and so: bright or stupid, friendly or unpleasant, talkative or shy, and so on. And so people are. But partly they are so because they were taught just as we were, because they feel that is the way a person should be. It does seem to simplify what might be confusing if everybody kept changing all the time. But, then, of course, that is just what everybody does do. Everybody *is* changing all the time. Just as the physiologists tell us all cells in our bodies are replaced each seven years or so, so the psychological aspects of us are in some change at all times.[51]

....

. . .What am I? The essence of my being cannot be pinned to any one palpable extension of my being. I am the process of I. I am the I-ing! - the ongoingness, the essence expressed in my existence. Not with Descartes' self-as-object inferring, "I doubt; hence I think. I think; hence I am." Rather, partaking of the very process of creation, "I am that I am!"[52]

Definitions of I (I process), Me and Self
. . .The *I*, as I conceive it, is irreducibly a unity and invariably a subject. It is, I postulate, the essential being.

Me is a word used to designate a perceptual object. The *Me* includes the physical body, customary patterns of behavior as an observer might note them, the memory of past actions and feelings. The *Me* of itself is inert, unaware, and it cannot make any of the defining statements the *I* makes.

Self is a word used with some overlapping of meaning with *Me*. The common element abstracted out of many and diverse perceptions of one's *Me* may be named the *Self* (just as we may

---

[51] Bugental, *op. cit.*, p. 393.
[52] *Ibid.*, p. 394.

speak of "tableness" from our experiences of many tables). This makes its significance synonymous with the *Self-concept*, and this will generally be my usage. . .

Person is a word we use to refer to another individual in his composite being, usually with emphasis on his inferred but unobservable *I*. In our discussions I will use the term *person* to speak about individuals when I do not want to specify whether I particularly refer to his *I*, *Me*, or *Self*.

The difficult central point of what I am trying to describe here is the concept of the pure subjectness of the *I*. Our habits of thinking are so tied up with the physical world and our related patterns of speech are so mechanically derived that we tend to think of entities as having both subject and object aspects always. Unthinkingly we then carry this over to our descriptions of the human experience. In a truly existential-humanistic framework, these habits lead to a major fallacy in conceiving our being. To set the matter at its most extreme: Only the *I* is truly the subject in world, whether the *I* be the *I* of a given *person* or the *I* which is all awareness. All other entities are objects.

This point is of crucial importance because the present confusion results in the *person* losing his sense of being, which means his awareness as the subject, the *I*, and comes to experience himself solely as object (Self, Me) and thus without genuine awareness, choice, relation, and so forth. From this confusion flow the sense of hollowness so endemic today, the kinds of contacts between people that are reminiscent of billiard balls bumping rather than beings in relation, the succumbing to absurdity and inertia, and much else that bespeaks our alienation from our intuition of being subject in our own lives.[53]

It does not seem possible to me to communicate adequately about the *I* while preserving faithful recognition of its pure subjectness and avoiding gramatical chaos. (This is particularly so because I use this same first person singular pronoun as the writer.) It may well be that this difficulty is pertinent and represents not alone a defect of language but of conceptualization. However that may be, I propose the convention

---

[53] Bugental, *op. cit.*, p. 201.

of referring to the *I*-process as a way of treating the *I* as though it were at times an object for our present communicative purposes.

What, then, is this *I* or *I-process*? In describing the given characteristics of being in Chapter 3 I identified the basic process as *feelingful awareness*. Thus, most basically, the I is that feelingful awareness. But, as I pointed out in my earlier discussion, our awareness discloses other aspects of our being: we are finite; we have the potential to take action; we can choose from among actions and non-actions that which we will make actual; and we are separate but related with other men. These, clearly, are the attributes of the *I-process*. The being of the individual person is what we point to by the term *I* or *I-process*. It is possible, for discussion purposes, to think of the *I-process* chiefly as combined awareness and choice-making, with the other attributes subsumed under this composite function... Now perhaps it will be evident from what has just been said that if the *I process* is conceived in this fashion, then such concepts as "health," "adjustment," "good," and "bad" are not pertinent. The *I-process* just *is*. The *I-process* is the livingness, but not just biological livingness. "Soul" or "essence" may be other recognitions of what is here termed the *I-process*. It is impossible to say at this point whether the *I-process* is an immortal, a metaphysical, or simply a bio-electro-chemical phenomenon. We can only describe what appears to be the nature of being and recognize that any speculation as to the deeper meaning of these observations must depend on further inquires.[54]

---

[54] Bugental, *op. cit.*, p. 204.

TABLE 1 BUGENTAL'S VIEW OF MAN : CONTENT ANALYSIS

No.	p. 41	Pp. 42-43	p. 44	pp. 45-46	pp. 46-49	pp. 49-51	pp. 51-33	First Sub-Total
1.		1	1		1	1		4
2.				2	2			4
3.						1		1
4.		1	3	2	1	1	1	9
5.								0
6.								0
7.								0
8.								0
9.		1					1	2
10.	1	1						2
11.								0
12.							1	1
13.								0
14.			1		1			2
15.						1		1
16.								0
17.								0
18.						1		1
19.								0
20.						1		1
21.								0
22.						1	1	2
23.								0
24.		1	2		2	2	1	8
25							1	1
26.								0
27.			1		1			2
28.								0
29.								0
30a.			1			1		2
30b.			1					1
				First cumulative	Sub-total			44

NOTE : The concept of 'authentic' has been coded as : aware plus self-determining plus ontologically and socially responsible.

Table 2 BUGENTAL'S VIEW OF MAN : CONTENT ANALYSIS

No.	pp. 54-55	pp. 56-57	pp. 58-61	pp. 62-63	pp. 64-65	Second Sub-Total	Total
1.			1			1	5
2.		3	1	1		5	9
3.					3	3	4
4.	2		1	1	1	5	14
5.			2			2	2
6.						0	0
7.						0	0
8.						0	0
9.						0	2
10.	1				1	2	4
11.					1	1	1
12.	1					1	2
13.						0	0
14.			1	1		2	4
15.						0	1
16.			1	1		2	2
17.				1		1	1
18.			1	1		2	3
19.						0	0
20.		3				3	4
21.						0	0
22.	2				1	3	5
23.			3			3	3
24.	1		3	1	2	7	15
25						0	1
26.						0	0
27.			1	1		2	4
28.						0	0
29.	1					1	1
30a.				1		1	3
30b.			1			1	2
				Second cumulative	Sub-total and cumulative total	48	92

Table 3 BUGENTAL'S VIEW OF MAN : PERCENTAGE EMPHASIS

No.	Dimension/Category	Total	Percentage Degree of Emphasis
1.	A communicative being	5	5.4
2.	Accepting of non-hedonic emotions	9	9.8
3.	Always in process	4	4.3
4.	Conscious/Aware	14	15.2
5.	Creative	2	2.2
6.	Forward-thrusting	0	0
7.	Freedom-cherishing	0	0
8.	Future-imagining	0	0
9.	Goal-directed	2	2.2
10.	Holistic/An integrated whole	4	4.3
11.	Intelligent	1	1.1
12.	Much like fellow man	2	2.2
13.	Nature-appreciating	0	0
14.	Ontologically responsible	4	4.3
15.	Open to life/experience	1	1.1
16.	Other-affirming	2	2.2
17.	Pleasure-loving	1	1.1
18.	Present-confronting	3	3.3
19.	Rational	0	0
20.	Risk-taking/Courageous	4	4.3
21.	Self-active	0	0
22.	Self-actualizing/Self-realizing	5	5.4
23.	Self-affirming	3	3.3
24.	Self-determining	15	16.3
25.	Self-disciplining	1	1.1
26.	Sexual	0	0
27.	Socially responsible	4	4.3
28.	Ultimately unknowable	0	0
29.	Unique	1	1.1
30a.	Reality-accepting	3	3.3
30b.	Ultimately alone	2	2.2
		92	100.0

SUMMARY

This chapter has consisted of the summarized version of Bugental's view of man and its analysis by means of the specially designed content analysis instrument.

In his view of man, Bugental has given primary emphasis (10 percent or more emphasis) to the following dimensions:

Man is self-determining

Man is conscious/aware

The areas of secondary emphasis (between 5 and 10 percent emphasis) were :

Man is accepting of non-hedonic emotions

Man is a communicative being

Man is self-actualizing

The remaining dimensions received tertiary emphasis (less than 5 percent emphasis).

Like Bonner, Bugental gives primary emphasis to the concept that man is self-determining. At each moment, man is faced with choices, and being endowed with the freedom to choose he is responsible for both the personal and social consequences of his choices.

Unlike Bonner, however, Bugental gives primary emphasis to the fact that man is conscious/aware. He sees feelingful awareness as the central process of man's nature and being as fully aware as possible the prime requisite for authentically confronting the basic existential anxieties and being self-actualizing in his choices rather than self-destructive. For Bugental, man is guided in his choices toward authenticity (this may also be termed 'genuine ontologic freedom' or 'full self-actualization') by his feelingful awareness and not by any built-in instincts. For Bonner, man makes his progress toward full self-actualization as a result of his natural yet socio-culturally influenced pro-action.

One final point needs to be pointed out. In the content analysis, Bugental's concept of authenticity which incorporates the idea of full self-actualization, has been coded as aware plus self-determining plus ontologically and socially responsible. The former two dimensions each received primary emphasis in the final count. Thus, Bugental has actually given primary emphasis to the dimension 'man is self-actualizing' although the percentage analysis table reveals only secondary emphasis for it. In the summaries that follow each chapter, the assumption has therefore been made that Bugental has given primary emphasis to the concept that man is a self-actualizing being.

6

ERICH FROMM'S VIEW OF MAN

Fromm's view of man is presented below in two parts. Part one consists of selected comments from his book, *Man for Himself* while part two is his credo as given in Chapter XII of his book, *Beyond the Chains of Illusion*.

PART ONE

...modern man feels uneasy and more and more bewildered. He works and strives, but he is dimly aware of a sense of futility with regard to his activities. While his power over matter grows, he feels powerless in his individual life and in society. While creating new and better *means* for mastering nature, he has become enmeshed in a network of those means and has lost the vision of the end which alone gives them significance-*man himself*. While becoming the master of nature, he has become the slave of the machine which his own hands built. With all his knowledge about matter, he is ignorant with regard to the most impotant and fundamental questions of human existence: what man is, how he ought to live, and how the tremendous energies *within* man can be released and used productively.[1]

. . .The progress of psychology lies not in the direction of divorcing an alleged "natural" from an alleged "spiritual" realm and focussing attention on the former, but in the return to the great tradition of humanistic ethics which looked

---

[1] Erich Fromm, Man for Himself (Greenwich), Conn. : Fawcett Publications Inc., 1947), p. 14.

at man in his physico-spiritual totality, believing that man's aim is *to be himself* and that the condition for attaining this goal is that man be *for himself.*

I have written this book with the intention of reaffirming the validity of humanistic ethics, to show that our knowledge of human nature does not lead to ethical relativism but, on the contrary, to the conviction that the sources of norms for ethical conduct are to be found in man's nature itself; that moral norms are based upon man's inherent qualities, and that their violation results in mental and emotional disintegration. I shall attempt to show that the character structure of the mature and integrated personality, the productive character, constitutes the source and the basis of "virtue" and that "vice," in the last analysis, is indifference to one's own self and self-mutilation. Not self-renunciation nor selfishness but self-love, not the negation of the individual but the affirmation of his truly human self, are the supreme values of humanistic ethics. If man is to have confidence in values, he must know himself and the capacity of his nature for goodness and productiveness.[2]

....

Authoritarian ethics can be distinguished from humanistic ethics by two criteria, one formal, the other material. Formally, authoritarian ethics denies man's capacity to know what is good or bad; the norm giver is always an authority transcending the individual... Materially, or according to content, authoritarian ethics answers the question of what is good or bad primarily in terms of the interests of the authority, not the interests of the subject; it is exploitative, although the subject may derive considerable benefits, psychic or material, from it.[3]

....

Humanistic ethics, in contrast to authoritarian ethics, may likewise be distinguished by formal and material criteria. *Formally,* it is based on the principle that only man himself can determine the criterion for virtue and aim, and not an authority transcending him. *Materially,* it is based on the principle that "good" is what is good for man and "evil"

---

[2] Fromm, op. cit., p.2.
[3] Ibid., p.20.

what is detrimental to man; *the sole criterion of ethical value being man's welfare.*[4]

....

Humanistic ethics is anthropocentric; not, of course, in the sense that man is the center of the universe but in the sense that his value judgements, like all other judgements and even perceptions, are rooted in the peculiarities of his existence and are meaningful only with reference to it; man, indeed, is the "measure of all things." The humanistic position is that there is nothing higher and nothing more dignified than human existence...[5]

....It is one of the characteristics of human nature that man finds his fulfillment and happiness only in relatedness to and solidarity with his fellow men. However, to love one's neighbor is not a phenomenon *transcending* man; it is something inherent in and *radiating from* him. Love is not a higher power which descends upon man nor a duty which is imposed upon him; it is his own power by which he relates himself to the world and makes it truly his.[6]

....

. . .*living itself is an art*-in fact, the most important and at the same time the most difficult and complex art to be practiced by man. Its object is not this or that specialized performance, but the performance of living, the process of developing into that which one is potentially. In the art of living, *man is both the artist and the object of his art*; he is the sculptor *and* the marble, the physician *and* the patient.[7]

....

If ethics constitutes the body of norms for achieving excellence in performing the art of living, its most general principles must follow from the nature of life in general and of human existence in particular. In most general terms, the nature of all life is to preserve and affirm its own existence. All organisms have an inherent tendency to preserve their existence: it is from this fact that psychologists have postulated an "instinct" of self-preservation. The first "duty" of an organism is to be alive.

---

[4] Fromm, *op. cit.,* p. 22.
[5] Ibld., p.23.
[6] Ibid., p.23.
[7] Ibid., p.27.

"To be alive" is a dynamic, not a static, concept. *Existence and the unfolding of the specific powers of an organism are one and the same.* All organisms have an inherent tendency to actualize their specific potentialities. *The aim of man's life,* therefore, is to be understood as *the unfolding of his powers according to the laws of his nature.*

Man, however, does not exist "in general." While sharing the core of human qualities with all members of his species, he is always an individual, a unique entity, different from everybody else. He differs by his particular blending of character, temperament, talents, dispositions, just as he differs at his fingertips. He can affirm his human potentialities only by realizing his individuality. The duty to be alive is the same as the duty to become oneself, to develop into the individual one potentially is.

To sum up, *good in humanistic ethics is the affirmation of life, the unfolding of man's powers. Virtue is responsibility toward his own existence.* Evil constitutes the crippling of man's powers; *vice is irresponsibility toward himself.*

These are the first principles of an objectivistic humanistic ethics...[8]

Fromm then comments on human nature and character :

One individual represents the human race. He is one specific example of the human species. He is "he" and he is "all"; he is an individual with his peculiarities and in this sense unique, and at the same time he is representative of all characteristics of the human race. His individual personality is determined by the peculiarities of human existence common to all men. Hence the discussion of the human situation must precede that of personality.[9]

The first element which differentiates human from animal existence is a negative one: the relative absence in man of instinctive regulation in the process of adaptation to the surrounding world. . .

. . .The emergence of man can be defined as occurring at the point in the process of evolution where instinctive adaptation

---

[8] Fromm, *op. cit.,* pp. 28-29.
[9] *Ibid., p. 47.*

has reached its minimum. But he emerges with new qualities which differentiate him from the animal: his awareness of himself as a separate entity, his ability to remember the past, to visualize the future, and to denote objects and acts by symbols; his reason to conceive and understand the world; and his imagination through which he reaches far beyond the range of his senses. Man is the most helpless of all animals, but this very biological weakness is the basis for his strength, the prime cause for the development of his specifically human qualities.[10]

Self-awareness, reason and imagination have disrupted the "harmony" which characterizes animal existence. Their emergence has made man into an anomaly, into the freak of the universe. He is part of nature, subject to her physical laws and unable to change them, yet he transcends the rest of nature. He is set apart while being a part; he is homeless, yet chained to the home he shares with all creatures. Cast into this world at an accidental place and time, he is forced out of it, again accidentally, Being aware of himself, he realizes his powerlessness and the limitations of his existence. He visualizes his own end: death. Never is he free from the dichotomy of his existence: he cannot rid himself of his mind, even if he should want to; he cannot rid himself of his body as long as he is alive - and his body makes him want to be alive.

Reason, man's blessing, is also his curse; it forces him to cope everlastingly with the task of solving an insoluble dichotomy. Human existence is different in this respect from that of all other organisms; it is in a state of constant and unavoidable disequilibrium... Man is the only animal for whom his own existence is a problem which he has to solve and from which he cannot escape. He cannot go back to the prehuman state of harmony with nature; he must proceed to develop his reason until he becomes the master of nature, and of himself.

The emergence of reason has created a dichotomy within man which forces him to strive everlastingly for new solutions.

---

[10] Fromm, *op. cit.*, p. 48.

The dynamism of his history is intrinsic to the existence of reason which causes him to develop and, through it, to create a world of his own in which he can feel at home with himself and his fellow men. Every stage he reaches leaves him discontented and perplexed, and this very perplexity urges him to move toward new solutions. There is no innate "drive for progress" in man; it is the contradiction in his existence that makes him proceed on the way he set out. . .

This split in man's nature leads to dichotomies which I call existential because they are rooted in the very existence of man; they are contradictions which man cannot annul but to which he can react in various ways, relative to his character and his culture.

The most fundamental existential dichotomy is that between life and death. The fact that we have to die is unalterable for man. Man is aware of this fact, and this very awareness profoundly influences his life...

That man is mortal results in another dichotomy: while every human being is the bearer of all human potentialities, the short span of his life does not permit their full realization under even the most favorable circumstances... Man's life, beginning and ending at one accidental point in the evolutionary process of the race, conflicts tragically with the individual's claim for the realization of all his potentialities. Of this contradiction between what he *could* realize and what he actually does realize he has, at least, a dim perception...

Man is alone and he is related at the same time. He is alone in as much as he is a unique entity, not identical with anyone else, and aware of his self as a separate entity. He must be alone when he has to judge or to make decisions solely by the power of his reason. And yet he cannot bear to be alone, to be unrelated to his fellow men. His happiness depends on the solidarity he feels with his fellow men, with past and future generations.

. . .It is one of the peculiar qualities of the human mind that, when confronted with a contradiction, it cannot remain passive. It is set in motion with the aim of resolving the contradiction.

All human progress is due to this fact...[11]

. . .There is only one solution to his problem: to face the truth, to acknowledge his fundamental aloneness and solitude in a universe indifferent to his fate, to recognize that there is no power transcending him which can solve his problem for him. Man must accept the responsibility for himself and the fact that only by using his own powers can he give meaning to his life. But meaning does not imply certainty; indeed, the quest for certainty blocks the search for meaning. Uncertainty is the very condition to impel man to unfold his powers. If he faces the truth without panic he will recognize that *there is no meaning to life except the meaning man gives his life by the unfolding of his powers, by living productively;* and that only constant vigilance, activity, and effort can keep us from failing in the one task that matters - the full development of our powers within the limitations set by the laws of our existence. Man will never cease to be perplexed, to wonder, and to raise new questions. Only if he recognizes the human situation, the dichotomies inherent in his existence and his capacity to unfold his powers, will he be able to succeed in his task: to be himself and for himself and to achieve happiness by the full realization of those faculties which are peculiarly his - of reason, love and productive work.[12]

. . .Fromm then proceeds to describe the productive character. Such a personality orientation, Fromm thinks, is within the reach of every normal human being. Thus, he writes :

. . .The "productive orientation" of personality refers to a fundamental attitude, a *mode of relatedness* in all realms of human experience. It covers mental, emotional, and sensory responses to others, to oneself, and to things. Productiveness is man's ability to use his powers and to realize the potentialities inherent in him. If we say he must use his powers

---

[11] Fromm, *op. cit.,* p. 48-52.
[12] *Ibid., p. 53.*

we imply that he must be free and not dependent on someone who controls his powers. We imply, furthermore, that he is guided by reason, since he can make use of his powers only if he knows what they are, how to use them, and what to use them for. Productiveness means that he experiences himself as the embodiment of his powers and as the "actor"; that he feels himself one with his powers and at the same time that they are not masked and alienated from him.[13]

. . .Productiveness is an attitude which every human being is capable of unless he is mentally and emotionally crippled.[14]

. . .In the concept of productiveness we are not concerned with activity *necessarily* leading to practical results but with an attitude, with a mode of reaction and orientation toward the world and oneself in the process of living. We are concerned with *man's character, not with his success.*

Productiveness is man's realization of the potentialities characteristic of him, the use of his *powers.* But what is "power"? It is rather ironical that this word denotes two contradictory concepts: *power of* — capacity and *power over* — domination. This contradiction, however, is of a particular kind…. *"Power over" is the perversion of "power to."* The ability of man to make productive use of his powers is his potency; the inability is his impotence. With his power of reason he can penetrate the surface of phenomena and understand their essence. With his power of love he can break through the wall which separates one person from another. With his power of imagination he can visualize things not yet existing; he can plan and thus begin to create. Where potency is lacking, man's relatedness to the world is perverted into the desire to dominate, to exert power over others as though they were things. Domination is coupled with death, potency with life. Domination springs from impotence and in turn reinforces it, for if an individual can force somebody else to serve him, his own

---

[13] Fromm, *op. cit.,* p. 91.
[14] *Ibid.,* p. 92.

need to be productive is increasingly paralysed.[15]

....

. . .While it is true that man's productiveness can create material things, works of art, and systems of thought, *by far the most important object of productiveness is man himself.* Birth is only one particular step in a continuum which begins with conception and ends with death. All that is between these two poles is a process of giving birth to one's potentialities, of bringing to life all that is potentially given in the two cells. But while physical growth proceeds by itself, if only the proper conditions are given, the process of birth on the mental plane, in contrast, does not occur automatically. It requires productive activity to give life to the emotional and intellectual potentialities of man, to give birth to his self. It is part of the tragedy of the human situation that the development of the self is never completed; even under the best conditions only part of man's potentialities is realized. Man always dies before he is fully born.[16]

In conclusion, we note the following passage :

Man's main task in life is to give birth to himself, to become what he potentially is. The most important product of his effort is his own personality. One can judge objectively to what extent the person has succeeded in his task, to what degree he has realized his potentialities. If he failed in his task, one can recognize this failure, and judge it for what it is—his moral failure. Even if one knows that the odds against the person were overwhelming and that everyone else would have failed too, the judgement about him remains the same. If one fully understands all the circumstances which made him as he is, one may have compassion for him; yet this compassion does not alter the validity of the judgement. Understanding a person does not mean condoning; it only means that one does not accuse him as if one were God or a judge placed above him.[17]

---

[15] Fromm, *op. cit.,* pp. 94-95.

[16] *Ibid., pp.* 97-98.

[17] *Ibid., p. 238.*

PART TWO

I believe that man is the product of natural evolution; that he is part of nature and yet transcends it, being endowed with reason and self-awareness.

I believe that man's essence is ascertainable. However, this essence is not a substance which characterizes man at all times through history. The essence of man consists in the above-mentioned contradiction inherent in his existence, and this contradiction forces him to react in order to find a solution. Man cannot remain neutral and passive toward this existential dichotomy. By the very fact of his being human, he is asked a question by life: how to overcome the split between himself and the world outside of him in order to arrive at the experience of unity and oneness with his fellow man and with nature. Man has to answer this question every moment of his life. Not only - or even primarily -with thoughts and words, but by his mode of being and acting.

I believe that there are a number of limited and ascertainable answers to this question of existence (the history of religion and philosophy is a catalogue of those answers); yet there are basically only two categories of answers. In one, man attempts to find again harmony with nature by regression to a prehuman from of existence, eliminating his specifically human qualities of reason and love. In the other, his goal is the full development of his human powers until he reaches a new harmony with his fellow man and with nature.

I believe that the first answer is bound to failure. It leads to death, destruction, suffering, and never to the full growth of man, never to harmony and strength. The second answer requires the elimination of greed and egocentricity, it demands discipline, will, and respect for those who can show the way. Yet, although this answer is the more difficult one, it is the only answer which is not doomed to failure. In fact, even before the final goal is reached, the activity and effort expended in approaching it has a unifying and integrating effect which intensifies man's vital energies

I believe that man's basic alternative is the choice between life and death. Every act implies this choice. Man is free to make it, but this freedom is a limited one. There are many favorable and unfavorable conditions which incline him — his psychological constitution, the condition of the specific society into which he was born, his family, teachers, and the friends he meets and chooses. It is man's task to enlarge the margin of freedom, to strengthen the conditions which are conducive to life as against those which are conducive to death. Life and death, as spoken of here, are not the biological states, but states of being, of relating to the world. Life means constant change, constant birth. Death means cessation of growth, ossification, repetition. The unhappy fate of many is that they do not make the choice. They are neither alive nor dead. Life become a burden, an aimless enterprise, and busyness is the means to protect one from the torture of being in the land of shadows.

I believe that neither life nor history has an ultimate meaning which in turn imparts meaning to the life of the individual or justifies his suffering. Considering the contradictions and weaknesses which beset man's existence it is only too natural that he seeks for an absolute which gives him the illusion of certainty and relieves him from conflict, doubt and responsibility. Yet, no god, neither in theological, philosophical or historical garments saves, or condemns man. Only man can find a goal for life and the means for the realization of this goal. He cannot find a saving ultimate or absolute answer but he can strive for a degree of intensity, depth and clarity of experience which gives him the strength to live without illusions, and to be free.

I believe that no one can "save" his fellow man by making the choice for him. All that one man can do for another is to show him the alternatives truthfully and lovingly, yet without sentimentality or illusion. Confrontation with the true alternatives may awaken all the hidden energies in a person, and enable him to choose life as against death. If he cannot choose life, no one else can breathe life into him.

I believe that there are two ways of arriving at the choice of the good. The first is that of duty and obedience to moral commands. This way can be effective, yet one must consider that in thousands of years only a minority have fulfilled even the requirements of the Ten Commandments. Many more have committed crimes when they were presented to them as commands by those in authority. The other way is to develop a taste for and a sense of well-being in doing what is good or right. By taste for well-being, I do not mean pleasure in the Benthamian or Freudian sense. I refer to the sense of heightened aliveness in which I confirm my powers and my identity.

I believe that education means to acquaint the young with the best heritage of the human race. But while much of this heritage is expressed in words, it is effective only if these words become reality in the person of the teacher and in the practice and structure of society. Only the idea which has materialized in the flesh can influence man; the idea which remains a word only changes words.

I believe in the perfectibility of man. This perfectibility means that man *can* reach his goal, but it does not mean that he *must* reach it. If the individual will not choose life and does not grow, he will by necessity become destructive, a living corpse. Evilness and self-loss are as real as are goodness and aliveness. They are the secondary potentialities of man if he chooses not to realize his primary potentialities.

I believe that only exceptionally is a man born as a saint or as a criminal. Most of us have dispositions for good and for evil, although the respective weight of these dispositions varies with individuals. Hence, our fate is largely determined by those influences which mold and form the given dispositions. The family is the most important influence. But the family itself is mainly an agent of society, the transmission belt for those values and norms which a society wants to impress on its members. Hence, the most important factor for the development of the individual is the structure and the values of the society into which he has been born.

I believe that society has both a furthering and an inhibiting function. Only in cooperation with others, and in the process of work, does man develop his powers, only in the historical process does he create himself. But at the same time, most societies until now have served the aims of the few who wanted to use the many. Hence they had to use their power to stultify and intimidate the many (and thus, indirectly, themselves), to prevent them from developing all their powers; for this reason society has always conflicted with humanity, with the universal norms valid for every man. Only when society's aim will have become identical with the aims of humanity, will society cease to cripple man and to further evil. I believe that every man represents humanity. We are different as to intelligence, health, talents. Yet we are all one. We are all saints and sinners, adults and children, and no one is anybody's superior or judge. We have all been awakened with the Buddha, we have all been crucified with Christ, and we have all killed and robbed with Genghis Khan, Stalin and Hitler.

I believe that man can visualize the experience of the whole universal man only by realizing his individuality and never by trying to reduce himself to an abstract, common denominator. Man's task in life is precisely the paradoxical one of realizing his individuality and at the same time transcending it and arriving at the experience of universality. Only the fully developed individual self can drop the ego.

I believe that the One World which is emerging can come into existence only if a New Man comes into being—a man who has emerged from the archaic ties of blood and soil, and who feels himself to be the son of man, a citizen of the world whose loyalty is to the human race and to life, rather than to any exclusive part of it; a man who loves his country because he loves mankind, and whose judgement is not warped by tribal loyalties.

I believe that man's growth is a process of continuous birth, of continuous awakening. We are usually half-asleep and only sufficiently awake to go about our business; but we are not awake enough to go about living, which is the only task that matters for a living being. The great leaders of the human race are those who have awakened man from his half-slumber. The great enemies of humanity are those who put it to sleep, and it does not matter whether their sleeping potion is the worship of God or that of the Golden Calf.

I believe that the development of man in the last four thousand years of history is truly awe-inspiring. He has developed his reason to a point where he is solving the riddles of nature, and has emancipated himself from the blind power of the natural forces. But at the very moment of his greatest triumph, when he is at the threshold of a new world, he has succumbed to the power of the very things and organizations he has created. He has invented a new method of producing, and has made production and distribution his new idol. He worships the work of his hands and has reduced himself to being the servant of things. He uses the name of God, of freedom, of humanity, of socialism, in vain; he prides himself on his powers—the bombs and the machines—to cover up his human bankruptcy; he boasts of his power to destroy in order to hide his human impotence.

I believe that the only force that can save us from self-destruction is reason; the capacity to recognize the unreality of most of the ideas that man holds, and to penetrate to the reality veiled by the layers and layers of deception and ideologies; reason, not as a body of knowledge, but as a "kind of energy, a force which is fully comprehensible only in its agency and effects. . ." a force whose "most important function consists in its power to bind and to dissolve." Violence and arms will not save us; sanity and reason may.

I believe that reason cannot be effective unless man has hope and belief. Goethe was right when he said that the deepest distinction between various historical periods is that between belief and disbelief, and when he added that all epochs in which belief dominates are brilliant, uplifting, and fruitful, while those in which disbelief dominates vanish because nobody cares to devote himself to the unfruitful. No doubt the thirteenth century, the Renaissance, the Enlightenment, were ages of belief and hope. I am afraid that the Western World in the twentieth century deceives itself about the fact that it has lost hope and belief. Truly, where there is no belief in man, the belief in machines will not save us from vanishing; on the contrary, this "belief" will only accelerate the end. Either the Western World will be capable of creating a renaissance of humanism in which the fullest development of man's humanity, and not production and work, are the central issues-or the West will perish as many other great civilizations have. I believe that to recognize the truth is not primarily a matter of intelligence, but a matter of character. The most important element is the courage to say *no*, to disobey the commands of power and of public opinion; to cease being asleep and to become human; to wake up and lose the sense of helplessness and futility. Eve and Prometheus are the two great rebels whose very "crimes" liberated mankind. But the capacity to say "no" meaningfully, implies the capacity to say "yes" meaningfully. The "yes" to God is the "no" to Caesar; the "yes" to man is the "no" to all those who want to enslave, exploit, and stultify him.

I believe in freedom, in man's right to be himself, to assert himself and to fight all those who try to prevent him from being himself. But freedom is more than the absence of violent oppression. It is more than "freedom from." It is "freedom to" - the freedom to become independent; the freedom to *be* much, rather than to *have* much, or to *use* things and people.

I believe that neither Western capitalism nor Soviet or Chinese communism can solve the problem of the future. They both create bureaucracies which transform man into a thing. Man must bring the forces of nature and of society under his conscious and rational control; but not under the control of a bureaucracy which administers things *and* man, but under the control of the free and associated producers who administer things and subordinate them to man, who is the measure of all things. The alternative is not between "capitalism" and "communism" but between bureaucratism and humanism. Democratic, decentralizing socialism is the realization of those conditions which are necessary to make the unfolding of all of man's powers the ultimate purpose.

I believe that one of the most disastrous mistakes in individual and social life consists in being caught in stereotyped alternatives of thinking. "Better dead than red," "an alienated industrial civilization or individualistic preindustrial society," "to rearm or to be helpless," are examples of such alternatives. There are always other and new possibilities which become apparent only when one has liberated oneself from the deathly grip of cliches, and when one permits the voice of humanity, and reason, to be heard. The principle of "the lesser evil" is the principle of despair. Most of the time it only lengthens the period until the greater evil wins out. To risk doing what is right and human, and have faith in the power of the voice of humanity and truth, is more realistic than the so-called realism of opportunism.

I believe that man must get rid of illusions that enslave and paralyze him; that he must become aware of the reality inside and outside of him in order to create a world which needs no illusions. Freedom and independence can be achieved only when the chains of illusion are broken.

I believe that today there is only one main concern: the question of war and peace. Man is likely to destroy all life on earth, or to destroy civilized life and the values among those that remain, and to build a barbaric, totalitarian organization which will rule what is left of mankind. To wake up to this danger, to look through the double talk on all sides which is used to prevent men from seeing the abyss toward which they are moving is the one obligation, the one moral and intellectual command which man must respect today. If he does not, we all will be doomed.

If we should all perish in the nuclear holocaust, it will not be because man was not capable of becoming human, or that he was inherently evil; it would be because the consensus of stupidity has prevented him from seeing reality and acting upon the truth.

I believe in the perfectibility of man, but I doubt whether he will achieve this goal, unless he awakens soon.

TABLE 1 FROMM'S VIEW OF MAN: CONTENT ANALYSIS

No.	pp. 73-74	pp. 75-76	pp. 76-79	pp. 80-81	First Sub-Total
1.		1	3		4
2.					0
3.					0
4.	1		4		5
5.	1			2	3
6.					0
7.					0
8.			1		1
9.					0
10.	1				1
11.			2	1	3
12.			1		1
13.					0
14.		1	1	1	3
15.					0
16.		1	1	1	3
17.					0
18.					0
19.			5	2	7
20.					0
21.			1		1
22.	1	6	5	7	19
23.	4	2	2	2	10
24.		1		2	3
25			1		1
26.					0
27.					0
28.					0
29.		1	3		4
30a.			1		1
30b.			4		4
				First cumulative Sub-total	74

Notes : The concept of 'productiveness' has been coded as: creative plus self-actualizing.

The concept of 'love' been coded as: self-affirming plus other affirming.

The concept of 'imagination' has been coded under the dimension 'intelligence.'

TABLE 2 FROMM'S VIEW OF MAN : CONTENT ANALYSIS

No.	pp. 82-83	pp. 84-85	pp. 85-86	pp. 87-88	Second Sub-Total	Total
1.		1			1	5
2.					0	0
3.	1				1	1
4.	1		2	1	4	9
5.					0	3
6.					0	0
7.	1				1	1
8.					0	1
9.	1				1	1
10.					0	1
11.					0	3
12.		1			1	2
13.					0	0
14.					0	3
15.					0	0
16.	1				1	4
17.					0	0
18.					0	0
19.	1		4	2	7	14
20.					0	0
21.					0	1
22.	1	4	1	2	8	27
23.		1		1	2	12
24.	3	1		1	5	8
25	1				1	2
26.					0	0
27.					0	0
28.					0	0
29.					0	4
30a.	2			1	3	4
30b.		1			1	5
			Second cumulative	Sub-total and cumulative total	37	111

Table 3 FROMM'S VIEW OF MAN : PERCENTAGE EMPHASIS

No.	Dimension/Category	Total	Percentage Degree of Emphasis
1.	A communicative being	5	4-5
2.	Accepting of non-hedonic emotions	0	0
3.	Always in process	1	0.9
4.	Conscious/Aware	9	8.1
5.	Creative	3	2.7
6.	Forward-thrusting	0	0
7.	Freedom-cherishing	1	0.9
8.	Future-imagining	1	0.9
9.	Goal-directed	1	0.9
10.	Holistic/An integrated whole	1	0.9
11.	Intelligent	3	2.7
12.	Much like fellow man	2	1.8
13.	Nature-appreciating	0	0
14.	Ontologically responsible	3	2.7
15.	Open to life/experience	0	0
16.	Other-affirming	4	3.6
17.	Pleasure-loving	0	0
18.	Present-confronting	0	0
19.	Rational	14	12.6
20.	Risk-taking/Courageous	0	0
21.	Self-active	1	0.9
22.	Self-actualizing/Self-realizing	27	24.3
23.	Self-affirming	12	10.8
24.	Self-determining	8	7.2
25.	Self-disciplining	2	1.8
26.	Sexual	0	0
27.	Socially responsible	0	0
28.	Ultimately unknowable	0	0
29.	Unique	4	3.6
30a.	Reality-accepting	4	3.6
30b.	Ultimately alone	5	4-5
		111	99.9*

*This amount differs from 100 percent because the figures were rounded off to the first decimal place for the sake of clarity.

SUMMARY

This chapter has consisted of the summarized version of Fromm's view of man and its analysis by means of the specially designed content analysis instrument.

In his view of man, Fromm has given primary emphasis (10 percent or more emphasis) to the following dimensions:

Man is self-actualizing

Man is rational

Man is self-affirming

The areas of secondary emphasis (between 5 and 10 percent emphasis) were:

Man is conscious/aware

Man is self-determining

The remaining dimensions received tertiary emphasis (less than 5 percent emphasis).

Like Bonner and Bugental, Fromm also considers man's full self-actulization or the unfolding and utilization of his powers and energies in accordance with the laws of his specific, idiosyncratic nature, to be the aim and meaning of his existence. He sees man as having this one over-riding need or value namely, to make himself into a productive personality, into a person who is able to love both himself and his fellow man.

However whereas Bonner stressed man's proaction, and Bugental man's feelingful awareness, Fromm emphasizes man's powers of reason as the guide for his choices in his quest for full self-actualization.

7

ABRAHAM H. MASLOW'S VIEW OF MAN

Maslow's view of man has been obtained from his book, *Motivation and Personality*. Relevant passages from this are quoted below to form a summarized version of this view.

...If I had had to condense the thesis of this book into a single sentence, I would have said that, in *addition* to what the psychologies of the time had to say about human nature, man also had a higher nature and that this was instinctoid, i.e., part of his essence. And if I could have had a second sentence, I would have stressed the profoundly holistic nature of human nature in contradiction to the analytic-dissecting-atomic-Newtonian approach of the behaviorisms and of Freudian psychoanalysis.[1]

....

I must say a word about the irriating fact that this veritable revolution (a new image of man, of society, of nature, of science, of ultimate values, of philosophy, etc., etc.) is still almost completely overlooked by much of the intellectual community, especially that portion of it that controls the channels of communication to the educated public and to youth. (For this reason I have taken to calling it the Unnoticed Revolution.)[2]

....

While it is still necessary to be very cautious about affirming the preconditions for "goodness" in human nature, it is already

---

[1] Abraham H. Maslow, *Motivation and Personality* (2d ed; New York. Harper and Row, 1970), p. ix.

[2] *Ibid., p.x.*

possible to reject firmly the despairing belief that human nature is ultimately and basically depraved and evil. Such a belief is no longer a matter of taste merely. It can now be maintained only by a determined blindness and ignorance, by a refusal to consider the facts..[3]

. . .The great advances of the last decade or so in the science of genetics has forced us to assign somewhat more determining power to the genes than we did fifteen years ago...[4]

....

. . .Currently, debate on the role of heredity and environment is almost as simplistic as it has been for the last fifty years. It still alternates between a simplistic theory of instincts on the one hand, total instincts of the sorts found in animals, and on the other hand, a complete rejection of the whole instinctual point of view in favor of a total environmentalism. Both positions are easily refuted, and in my opinion are so untenable as to be called stupid. In contrast to these polarized positions the theory set forth in Chapter 6 and throughout the remainder of the book gives a third position, namely, that there are *very weak* instinct-remnants left in the human species, nothing that could be called full instincts in the animal sense. These instinct-remnants and instinctoid tendencies are so weak that culture and learning easily overwhelm them and must be considered to be far more powerful. In fact, the techniques of psychoanalysis and other uncovering therapies, let alone the "quest for identity," may all be conceived as the very difficult and delicate task of discovering through the overlay of learning, habit, and culture, what our instinct-remnants and instinctoid tendencies, our weakly indicated essential nature may be. In a word, man has a biological essence, but this is very weakly and subtly determined, and needs special hunting techniques to discover it; we must discover, individually and subjectively, our animality, our specieshood.[5]

....

Ultimately this point of view will force us to take far more

---

[3] Maslow, *op. cit.*, pp. x-xi.
[4] *Ibid., p.* xvii.
[5] *Ibid., pp. xvii-xviii.*

seriously than we do the fact of individual differences, as well as species membership. We will have to learn to think of them in this new way as being, (1) very plastic, superficial, easily changed, easily stamped out, but producing thereby all sorts of subtle pathologies. This leads to the delicate task, (2) of trying to uncover the temperament, the constitution, the hidden *bent* of each individual so that he can grow unhampered in his own individual style. This attitude will require far greater attention than has been given by the psychologists to the subtle psychological and physiological costs and sufferings of denying one's true bent, sufferings that are not necessarily conscious or easily seen from the outside...[6]

Maslow then looks at the nature of man from the point of view of what motivates him.

Our first proposition states that the individual is an integrated, organized whole. This theoretical statement is usually accepted piously enough by psychologists, who then often proceed calmly to ignore it in their actual experiments. That it is an experimental reality as well as a theoretical one must be realized before sound experimentation and sound motivation theory are possible. In motivation theory this proposition means many specific things. For instance, it means the whole individual is motivated rather than just a part of him. In good theory there is no such entity as a need of the stomach or mouth, or a genital need. There is only a need of the individual. It is John Smith who wants food, not John Smith's stomach. Furthermore satisfaction comes to the whole individual and not just to a part of him. Food satisfies John Smith's hunger and not his stomach's hunger.

Dealing with John Smith's hunger as a function merely of his gastrointestinal tract has made experimenters neglect the fact that when an individual is hungry he changes not only in his gastrointestinal function, but in many, perhaps even in most other functions of which he is capable. His perceptions change

---

(he will perceive food more readily than he will at other times). His memories change (he is more apt to remember a good meal at this time than at other times). His emotions change (he is more tense and nervous than he is at other times). The content of his thinking changes (he is more apt to think of getting food than of solving an algebraic problem). And this list can be extended to almost every other faculty, capacity, or function, both physiological and psychic. In other words, when John Smith is hungry, he is hungry all over; he is different as an individual from what he is at other times.[7]

Man is a wanting animal and rarely reaches a state of complete satisfaction except for a short time. As one desire is satisfied, another pops up to take its place. When this is satisfied, still another comes into the foreground, etc. It is a characteristic of the human being throughout his whole life that he is practically always desiring something...[8]

Dewey and Thorndike have stressed one important aspect of motivation that has been completely neglected by most psychologists, namely, possibility. On the whole we yearn consciously for that which might conceivably be actually attained. That is to say that we are much more realistic about wishing than the psychoanalysts may allow, absorbed as they are with unconscious wishes.

As a man's income increases he finds himself actively wishing for and striving for things that he never dreamed of a few years before. The average American yearns for automobiles, refrigerators, and television sets because they are real possibilities; he does not yearn for yachts of planes because they are in fact not within the reach of the average American. It is quite probable that he does not long for them *unconsciously* either.

Attention to this factor of possibility of attainment is crucial for understanding the differences in motivations between various classes and castes within our own population and

---

[7] Maslow, *op cit.* p. 19.
[8] *Ibid.*, p. 24.

between it and other poorer countries and cultures.[9]

....

The needs that are usually taken as the starting point for motivation theory are the so-called physiological drives...[10]

....

Undoubtedly these physiological needs are the most prepotent of all needs. What this means specifically is that in the human being who is missing everything in life in an extreme fashion, it is most likely that the major motivation would be the physiological needs rather than any others A person who is lacking food, safety, love, and esteem would most probably hunger for food more strongly than for anything else.

If all the needs are unsatisfied, and the organism is then dominated by the physiological needs, all other needs may become simply nonexistent or be pushed into the background. It is then fair to characterize the whole organism by saying simply that it is hungry, for consciousness is almost completely preempted by hunger...[11]

....

. . .But what happens to man's desires when there is plenty of bread and when his belly is chronically filled?

At once other (and higher) needs emerge and these, rather than physiological hungers, dominate the organism. And when these in turn are satisfied, again new (and still higher) needs emerge, and so on. This is what we mean by saying that the basic human needs are organized into a hierarchy of relative prepotency.

. . .a want that is satisfied is no longer a want. The organism is dominated and its behavior organized only by unsatisfied needs....[12]

....

If the physiological needs are relatively well gratified, there then emerges a new set of needs, which we may characterize roughly as the safety needs (security; stability; dependency;

---

[9] Maslow, *op cit.* p. 31.
[10] *Ibid.*, p. 35.
[11] *Ibid.*, pp. 36-37.
[12] *Ibid.*, p. 38.

protection; freedom from fear, from anxiety and chaos; need for structure, order, law, limits; strength in the protector; and so on)...[13]

....

If both the physiological and the safety needs are fairly well gratified, there will emerge the love and affection and belongingness needs, and the whole cycle already described will repeat itself with this new center...[14]

....

All people in our society (with a few pathological exceptions) have a need or desire for a stable, firmly based, usually high evaluation of themselves, for self-respect, or self-esteem, and for the esteem of others. These needs may therefore be classifield into two subsidiary sets. These are, first, the desire for strength, for achievement, for adequacy, for mastery and competence, for confidence in the face of the world, and for independence and freedom. Second, we have what we may call the desire for reputation or prestige (defining it as respect or esteem from other people), status, fame and glory, dominance, recognition, attention, importance, dignity, or appreciation. These needs have been relatively stressed by Alfred Adler and his followers, and have been relatively neglected by Freud. More and more today, however, there is appearing widespread appreciation of their central importance...[15]

....

Even is all these needs are satisfied, we may still often (if not always) expect that a new discontent and restlessness will soon develop, unless the individual is doing what *he*, individually, is fitted for. A musician must make music, an artist must paint, a poet must write, if he is to be ultimately at peace with himself. What a man *can* be, he *must* be. He must be true to his own nature. This need we may call self-actualization...This term, first coined by Kurt Goldstein, is being used in this book in a much more specific and limited fashion. It refers to man's desire for self-fulfillment,   namely,   to   the   tendency   for

---

[13] Maslow, *op cit.* p. 39.
[14] *Ibid.*, p. 43.
[15] *Ibid., p. 45*

him to become actualized in what he is potentially. This tendency may be phrased as the desire to become more and more what one idiosyncratically is, to become everything that one is capable of becoming.

The specific form that these needs will take will of course vary greatly from person to person. In one individual it may take the form of the desire to be an ideal mother, in another it may be expressed athletically, and in still another it may be expressed athletically, and in still another it may be expressed in painting pictures or in inventions. At this level, individual differences are greatest.

The clear emergence of these needs usually rests upon some prior satisfaction of the physiological, safety, love, and esteem needs.[16]

The remaining portion of this summarized version of Maslow's view of man is based on his study of self-actualizing people. The subjects were selected from among personal acquaintances and friends, and among public and historical figures. Maslow explains the meaning of self-actualization :

. . . The positive criterion for selection was positive evidence of self-actualization (SA), as yet a difficult syndrome to describe accurately. For the purposes of this discussion, it may be loosely described as the full use and exploitation of talents, capacities, potentialities, etc. Such people seem to be fulfilling themselves and to be doing the best that they are capable of doing, reminding us of Nietzsche's exhortation, "Become what thou art!" They are people who have developed or are developing to the full stature of which they are capable...[17]

Since the need for self-actualization has been postulated as an instinctoid need in man, the characteristics of the self-actualizing people can be considered to be presently existing,

---

[16] Maslow, *op cit.* p. 46.

[17] *Ibid.,* p. 150.

though in the form of unrealized potentials, in man in general. Consequently, these are noted as part of Maslow's view of man:[18]

1. *More Efficient Perception of Reality and More Comfortable Relations with it*

The first form in which this capacity was noticed was an unusual ability to detect the spurious, the fake, and the dishonest in personality, and in general to judge people correctly and efficiently...

As the study progressed, it slowly become apparent that this efficiency extended to many other areas of life - indeed *all* areas that were observed. In art and music, in things of the intellect, in scientific matters, in politics and public affairs, they seemed as a group to be able to see concealed or confused realities more swiftly and more correctly than others...[19]

2. *Acceptance (self, other, nature)*

...Our healthy individuals find it possible to accept themselves and their own nature without chagrin or complaint or, for that matter, even without thinking about the matter very much.

They can accept their own human nature in the stoic style, with all its shortcomings, with all its discrepancies from the ideal image without feeling real concern. It would convey the wrong impression to say that they are self-satisfied. What we must say rather is that they can take the frailties and sins, weaknesses, and evils of human nature in the same unquestioning spirit with which one accepts the characteristics of nature. One does not complain about water because it is wet, or about rocks because they are hard, or about trees because they are green. As the child looks out upon the world with wide, uncritical, undemanding, innocent eyes, simply noting and observing what is the case, without either arguing the matter or demanding that it be otherwise, so does the

---

[18] Maslow, *op cit.* p. 153-174.
[19] *Ibid.*, p. 153.

self-actualizing person tend to look upon human nature in himself and in others...[20]

3. *Spontaneity; Simplicity; Naturalness*

Self-actualizing people can all be described as relatively spontaneous in behavior and far more spontaneous than that in their inner life, thoughts, impulses, etc. Their behavior is marked by simplicity and naturalness, and by lack of artificiality or straining for effect. This does not necessarily mean consistently unconventional behavior. If we were to take an actual count of the number of times that the self-actualizing person behaved in an unconventional manner the tally would not be high. His unconventionality is not superficial but essential or internal. It is his impulses, thought, consciousness that are so unusually unconventional, spontaneous, and natural...[21]

4. *Problem Centering*

Our subjects are in general strongly focussed on problems outside themselves. In current terminology they are problem centered rather than ego centered. They generally are not problems for themselves and are not generally much concerned about themselves; e.g., as contrasted with the ordinary introspectiveness that one finds in insecure people. These individuals customarily have some mission in life, some task to fulfill, some problem outside themselves which enlists much of their energies.[22]

5. *The Quality of Detachment; the Need for Privacy*

For all my subjects it is true that they can be solitary without harm to themselves and without discomfort. Furthermore, it is true for almost all that they positively *like* solitude and privacy to a definitely greater degree than the average person.[23]

---

[20] Maslow, *op cit.* p. 155.
[21] *Ibid.*, p. 157
[22] *Ibid.*, p. 159.
[23] *Ibid.*, p. 160.

. . . .

This quality of detachment may have some connection with certain other qualities as well. For one thing it is possible to call my subjects more objective (in *all* senses of that word) than average people. We have seen that they are more problem centered than ego centered. This is true even when the problems concern themselves, their own wishes, motives, hopes, or aspirations. Consequently, they have the ability to concentrate to a degree not usual for ordinary men. . .

. . . .

. . . .My subjects make up their own minds, come to their own decisions, are self-starters, are responsible for themselves and their own destinies...

. . . .Of my self-actualizing subjects, 100 percent are self-movers.[24]

Finally, I must make a statement even though it will certainly be disturbing to many theologians, philosophers, and scientists: self-actualizing individuals have more "free will" and are less "determined" than average people are...[25]

6 *Autonomy; Independence of Culture and Environment; Will; Active* Agents

One of the characteristics self-actualizing people, which to a certain extent crosscuts much of what we have already described, is their relative independence of the physical and social environment. Since they are propelled by growth motivation rather than by deficiency motivation, self-actualizing people are not dependent for their main satisfactions on the real world, or other people or culture or means to ends or, in general, on extrinsic satisfactions. Rather they are dependent for their own development and continued growth on their own potentialities and latent resources. Just as the tree needs sunshine and water and food, so do most people need love, safety, and the other basic need gratifications that can come only from without. But once these external satisfiers are obtained, once these inner deficiencies are satiated by outside

---

[24] Maslow, *op cit.* p. 160-161.

[25] *Ibid.,* pp. 161-162.

satisfiers, the true problem of individual human development begins, e.g., self-actualization.

This independence of environment means a relative stability in the face of hard knocks, blows, deprivations, frustrations, and the like. These people can maintain a relative serenity in the midst of circumstances that would drive other people to suicide; they have also been described as "self-contained."

Deficiency-motivated people *must* have other people available, since most of their main need gratifications (love, safety, respect, prestige, belongingness) can come only from other human beings. But growth-motivated people may actually be *hampered* by others. The determinants of satisfaction and of the good life are for them now inner-individual and *not* social. They have become strong enough to be independent of the good opinion of other people, or even of their affection...[26]

7. *Continued Freshness of Appreciation*

Self-actualizing people have the wonderful capacity to appreciate again and again, freshly and naively, the basic goods of life, with awe, pleasure, wonder, and even ecstasy, however stale these experiences may have become to others— what C. Wilson has called "newness." Thus for such a person, any sunset may be as beautiful as the first one, any flower may be of breath-taking loveliness, even after he has seen a million flowers. The thousandth baby he sees is just as miraculous a product as the first one he saw. He remains as convinced of his luck in marriage thirty years after his marriage and is as surprised by his wife's beauty when she is sixty as he was forty years before. For such people, even the casual workaday, moment-to-moment business of living can be thrilling, exciting, and ecstatic. These intense feelings do not come all the time; they come occasionally rather than usually, but at the most unexpected moments...[27]

---

[26] Maslow, *op.cit.*, p. 162
[27] Ibid., p. 163.

8. The Mystic Experience; the Peak Experience

Those subjective expressions that have been called the mystic experience and described so well by William James are a fairly common experience for our subjects though not for all. The strong emotions described in the previous section sometimes get strong enough, chaotic, and widespread enough to be called mystic experiences.....

....

Apparently the acute mystic or peak experience is a tremendous intensification of *any* of the experiences in which there is loss of self or transcendence of it, e.g., problem centering, intense concentration, muga behavior, as described by Benedict, intense sensuous experience, self-forgetful and intense enjoyment of music or art... [28]

9. Gemeinschaftsgefühl

This word, invented by Alfred Adler, is the only one available that describes well the flavor of the feelings for mankind expressed by self-actualizing subjects. They have for human beings in general a deep feeling of identification, sympathy, and affection in spite of the occasional anger, impatience, or disgust described below. Because of this they have a genuine desire to help the human race. It is as if they were all members of a single family...[29]

*10. Interpersonal Relations*SA

Self-actualizing people have deeper and more profound interpersonal relations than any other adults... They are capable of more fusion, greater love, more perfect identification, more obliteration of the ego boundaries than any other people would consider possible... One consequence of this phenomenon and of certain others as well is that self-actualizing people have these especially deep ties with rather few individuals. Their circle of friends is rather

---

[28] Maslow, op.cit., p. 164-165.
[29] Ibid., p. 165.

small. The ones that they love profoundly are few in number. Party this is for the reason that being very close to someone in this self-actualizing style seems to require a good deal of time...[30]

11. *The Democratic Character Structure*

All my subjects without exception may be said to be democratic people in the deepest possible sense. I say this on the basis of a previous analysis of authoritarian and democratic character structures. . . . These people have all the obvious or superficial democratic characteristics. They can be and are friendly with anyone of suitable character regardless of class, education, political belief, race, or color. As a matter of fact it often seems as if they are not even aware of these differences, which are for the average person so obvious and so important.

They have not only this most obvious quality but their democratic feeling goes deeper as well. For instance they find it possible to learn from anybody who has something to teach them — no matter what other characteristics he may have. . .

Most profound, but also most vague is the hard-to-get-at-tendency to give a certain quantum of respect to *any* human being just because he is a human individual; our subjects seem not to wish to go beyond a certain minimum point, even with scoundrels, of demeaning, of derogating, of robbing of dignity. And yet, this goes along with their strong sense of right and wrong, of good and evil. They are *more* likely rather than less likely to counterattack against evil men and evil behavior. They are far less ambivalent, confused or weak-willed about their own anger than average men are.[31]

12. *Discrimination between Means and Ends, between Good and Evil*

I have found none of my subjects to be chronically unsure about the difference between right and wrong in his actual

---

[30] Maslow, *op cit.* p. 166.
[31] *Ibid.,* p. 167.

living. Whether or not they could verbalize the matter, they rarely showed in their day-to-day living the chaos, the confusion, the inconsistency, or the conflict that are so common in the average person's ethical dealings. This may be phrased also in the following terms: these individuals are strongly ethical, they have definite moral standards, they do right and do not do wrong. Needless to say, their notions of right and wrong and of good and evil are often not the conventional ones.[32]

13. *Philosophical, Unhostile Sense of Humor*

One very early finding that was quite easy to make, because it was common to all of my subjects, was that their sense of humor is not of the ordinary type. They do not consider funny what the average man considers to be funny. Thus they do not laugh at hostile humor (making people laugh by hurting someone) or superiority humor (laughing at some one else's inferiority) or authority-rebellion humor (the unfunny, Oedipal, or smutty joke). Characteristically what they consider humor is more closely allied to philosophy than to anything else. . .[33]

14. *Creativeness$_{SA}$*

This is a universal characteristic of all the people studied or observed. There is no exception. Each one shows in one way or another a special kind of creativeness or originality or inventiveness that has certain peculiar characteristics. . . The creativeness of the self-actualized man seems. . . to be kin to the naive and universal creativeness of unspoiled children. It seems to be more a fundamental characteristic of common human nature—a potentiality given to all human beings at birth. Most human beings lose this as they become enculturated, but some few individuals seem either to retain this fresh and naive, direct way of looking at life, or if they have lost it, as most people do, they later in life recover it.

---

[32] Maslow, *op cit.* p. 168.
[33] *Ibid.,* p. 169.

Santayana called this the "second naivete," a very good name for it.[34]

15. *Resistance to Enculturation; the Transcendence of any Particular Culture*

Self-actualizing people are not well adjusted (in the naive sense of approval of and identification with the culture). They get along with the culture in various ways, but of all of them it may be said that in a certain profound and meaningful sense they resist enculturation and maintain a certain inner detachment from the culture in which they are immersed...[35]

16. *The Imperfections of Self-actualizing People*

Our subjects show many of the lesser human failings. They too are equipped with silly, wasteful, or thoughtless habits. They can be boring, stubborn, irritating. They are by no means free from a rather superficial vanity, pride, partiality to their own productions, family, friends, and children. Temper outbursts are not rare.

Our subjects are occasionally capable of an extraordinary and unexpected ruthlessness. It must be remembered that they are very strong people. This makes it possible for them to display a surgical coldness when this is called for, beyond the power of the average man...[36]

....

Finally, it has already been pointed out that these people are *not* free of guilt, anxiety, sadness, self-castigation, internal strife, and conflict. The fact that these arise out of non-neurotic sources is of little consequence to most people today (even to most psychologists) who are therefore apt to think them *un*healthy for this reason.

What this has taught me I think all of us had better learn. *There are no perfect human beings!* Persons can be found who

---

[34] Maslow, *op cit,* p. 170.
[35] *Ibid.,* p. 171.
[36] *Ibid.,* p. 175.

are good, very good indeed, in fact, great. There do in fact exist creators, seers, sages, saints, shakers and movers. This can certainly give us hope for the future of the species even if they *are* uncommon and do *not* come by the dozen. And yet these very same people can at times be boring, irritating, petulant, selfish, angry, or depressed. To avoid disillusionment with human nature, we must first give up our illusions about it.[37]

17. *Values and Self-actualization*

A firm foundation for a value system is automatically furnished to the self-actualizer by his philosophic acceptance of the nature of his self, of human nature, of much of social life, and of nature and physical reality. These acceptance values account for a high percentage of the total of his individual value judgements from day to day. What he approves of, disapproves of, is loyal to, opposes or proposes, what pleases him or displeases him can often be understood as surface derivations of this source trait of acceptance.[38]

....

The topmost portion of the value system of the Self-actualized person is entirely unique and idiosyncratic-character-structure expressive. This must be true by definition, for self-actualization is actualization of a self, and no two selves are altogether alike. There is only one Renoir, one Brahms, one Spinoza. Our subjects had very much in common, as we have seen and yet at the same time were more completely individualized, more unmistakably themselves, less easily confounded with others than any average control group could possibly be. That is to say, they are simultaneously very much alike and very much unlike each other. They are more completely individual than any group that has ever been described, and yet also more completely socialized, more identified with humanity than any other group yet described. They are closer to *both* their specieshood and to their unique individuality.[39]

---

[37] Maslow, *op cit.* p. 176.

[38] *Ibid.,* p. 176.

[39] *Ibid.,* p. 178.

18. *The Resolution of Dichotomies in Self-actualization*

At this point we may finally allow ourselves to generalize and underscore a very important theoretical conclusion derivable from the study of self-actualizing people. At several points in this chapter—and in other chapters as well—it was concluded that what had been considered in the past to be polarities or opposites or dichotomies were so *only in less healthy people*. In healthy people, these dichotomies were resolved, the polarities disappeared and many oppositions thought to be intrinsic merged and coalesced with each other to form unities. ..

For example the age-old opposition between heart and head, reason and instinct, or cognition and conation was seen to disappear in healthy people where they become synergic rather than antagonists, and where conflict between them disappears because they say the same thing and point to the same conclusion. In a word in these people, desires are in excellent accord with reason. St. Augustine's "Love God and do as you will" can easily be translated, "Be healthy and then you may trust your impulses."

The dichotomy between selfishness and unselfishness disappears altogether in healthy people because in principle every act is *both* selfish and unselfish. Our subjects are simultaneously very spiritual and very pagan and sensual even to the point where sexuality becomes a *path* to the spiritual and "religious". Duty cannot be contrasted with pleasure nor work with play when duty *is* pleasure, when work *is* play, and the person doing his duty and being virtuous is simultaneously seeking his pleasure and being happy. If the most socially identified people are themselves also the most individualistic people, of what use is it to retain the polarity? If the most mature are also childlike? And if the most ethical and moral people are also the lustiest and most animal?

Similar findings have been reached for kindness-ruthlessness, concreteness-abstractness, acceptance- rebellion, self-society, adjustment-maladjustment, detachment from others-identification with others, serious-humorous, Dionysian-Apollonian, introverted-extraverted, intense-casual, serious-frivolous, conventional-unconventional, mystic-realistic, active-passive, masculine-feminine, lust-love, and Eros-Agape. . .

In this, as in other ways, healthy people are so different from average ones, not only in degree but in kind as well, that they generate two very different kinds of psychology. It becomes more and more clear that the study of crippled, stunted, immature, and unhealthy specimens can yield only a cripple psychology and a cripple philosophy. The study of self-actualizing people must be the basis for a more universal science of psychology.[40]

Table1 MASLOWS VIEW OF MAN : CONTENT ANALYSIS

No.	pp. 93-94	pp. 95-99	pp. 100-103	pp. 104-106	pp. 107-110	Total
1.						0
2.					1	1
3.						0
4.						0
5.				1		1
6.						0
7.						0
8.						0
9.		1	1			2
10.	1	2			2	5
11.						0
12.					1	1
13.		1				1
14.				1		1
15.						0
16.			1	3	1	5
17.						0
18.						0
19.						0
20.						0
21.			1			1
22.		4	2		1	7
23.	1	1	2	1	1	6
24.			2			2
25						0
26.						0
27.				2		2
28.						0
29.	1	2			2	5
30a.		1	1			2
30b.						0
				Cumulative total		42

111

TABLE 2 MASLOWS VIEW OF MAN: PERCENTAGE EMPHASIS

No.	Dimension/Category	Total	Percentage Degree of Emphasis
1.	A communicative being	0	0
2.	Accepting of non-hedonic emotions	1	2.4
3.	Always in process	0	0
4.	Conscious/Aware	0	0
5.	Creative	1	2.4
6.	Forward-thrusting	0	0
7.	Freedom-cherishing	0	0
8.	Future-imagining	0	0
9.	Goal-directed	2	4.8
10.	Holistic/An integrated whole	5	12.0
11.	Intelligent	0	0
12.	Much like fellow man	1	2.4
13.	Nature-appreciating	1	2.4
14.	Ontologically responsible	1	2.4
15.	Open to life/experience	0	0
16.	Other-affirming	5	12.0
17.	Pleasure-loving	0	0
18.	Present-confronting	0	0
19.	Rational	0	0
20.	Risk-taking/Courageous	0	0
21.	Self-active	1	2.4
22.	Self-actualizing/Self-realizing	7	16.8
23.	Self-affirming	6	14.4
24.	Self-determining	2	4.8
25.	Self-disciplining	0	0
26.	Sexual	0	0
27.	Socially responsible	2	4.8
28.	Ultimately unknowable	0	0
29.	Unique	5	12.0
30a.	Reality-accepting	2	4.8
30b.	Ultimately alone	0	0
		42	100.8%*

*This amount differs from 100 percent because the figures were rounded off to the first decimal place for the sake of clarity.

SUMMARY

This chapter has consisted of the summarized version of Maslow's view of man and its analysis by means of the specially designed content analysis instrument.

In his view of man, Maslow has given primary emphasis (10 percent or more emphasis) to the following dimensions:

Man is self-actualizing

Man is self-affirming

Man is holistic/An integrated whole

Man is other-affirming

Man is unique

All the remaining dimensions received tertiary emphasis (less than 5 percent emphasis).

As in the case of Bonner, Bugental and Fromm, Maslow also gives primary emphasis to the concept that man is a self-actualizing being. Simultaneously, however, he gives primary emphasis to the concepts of 'self-affirming' and 'other-affirming.' It would appear that these three interrelated concepts can be brought under the one idea that man is synergic. In other words, man's nature is such that he values not only his own self-actualization but, simultaneously, he values the self-actualization of the other person. Man's choices can then no longer be labelled as selfish or unselfish because the synergic person transcends this dichotomy and aids his fellow man to achieve a fuller realization of humane values at the same time as he does this for himself. Thus, though Maslow explicitly points to selfactualization as an instinctoid need of man's deeper/higher nature, he could be better understood as saying that man has an instinctoid need for synergic actualization.

8

ROLLO MAY'S VIEW OF MAN

May's view of man has been obtained from his book, *Man's Search for Himself.* Relevant passages from this have been quoted below to form a summarized version of his view.

May first addresses himself to some of the problems that men experience.

It may sound surprising when I say, on the basis of my own clinical practice as well as that of my psychological and psychiatric colleagues, that the chief problem of people in the middle decade of the twentieth century is *emptiness*. By that I mean not only that many people do not know what they want; they often do not have any clear idea of what they feel...[1]

While one might laugh at the meaningless boredom of people a decade or two ago, the emptiness has for many now moved from the state of boredom to a state of futility and despair which holds promise of dangers. . . The human being cannot live in a condition of emptiness for very long: if he is not growing toward something, he does not merely stagnate; the pent-up potentialities turn into morbidity and despair, and eventually into destructive activities.[2]

Another characteristic of modern people is loneliness. They

---

[1] Rollo May, *Man's Search for Himself* (New York : Signet Books, 1953), pp. 13-14.

[2] *Ibid.,* p. 22.

describe this feeling as one of being "on the outside, "isolated, or, if they are sophisticated, they say that they feel alienated...[3]

Anxiety, the other characteristic of modern man, is even more basic than emptiness and loneliness. For being "hollow" and lonely would not bother us except that it makes us prey to that peculiar psychological pain and turmoil called anxiety.[4]

. . .anxiety may take all form and intensities, for it is the human being's basic *reaction to a danger to his existence, or to some value he identifies with his existence....* as soon as the threat becomes great enough to involve the total self, one then has the experience of anxiety. Anxiety strikes us at the very "core" of ourselves: it is what we feel when our existence as selves is threatened.

. . .In its full-blown intensity, anxiety is the most painful emotion to which the human animal is heir...[5]

Having said this, May goes on to elaborate on the two central themes of his view of man :

1. Man has consciousness (awareness) of himself, and he must use this.
2. Man also has other unique powers which too he must use to fulfill his own potentialities.

Thus, May notes :

. . .around the age of two, more or less, there appears in the human being the most radical and important emergence so far in evolution, namely his consciousness of himself. He begins to be aware of himself as an "I." As the foetus in the womb, the infant has been part of the "original we" with its mother,

---

[3] May, *op. cit.*, p. 23.
[4] *Ibid.*, p. 30.
[5] *Ibid.*, p. 35.

and it continues as part of the psychological "we" in early infancy. But now the little child–for the first time–becomes aware of his freedom. . . He experiences himself as an identity who is separated from his parents and can stand against them is need be. This remarkable emergence is the birth of the human animal into a person.[6]

This consciousness of self, this capacity to see one's self as though from the outside, is the distinctive characteristic of man.[7]

. . . .

. . .man's consciousness of himself is the source of his highest qualities...[8]

This capacity for consciousness of ourselves gives us the ability to see ourselves as others see us and to have empathy with others... No matter how poorly we use or fail to use or even, abuse these capacities, they are the rudiments of our ability to begin to love our neighbor, to have ethical sensitivity, to see truth, to create beauty, to devote ourselves to ideals, and to die for them if need be.

To fulfill these potentialities is to be a person...[9]

. . . .

... Almost every adult is, in greater or lesser degree, still struggling on the long journey to achieve selfhood on the basis of the patterns which were set in his early experiences in the family. Nor do we for a moment overlook the fact that selfhood is always born in a social context... Or, as we put it above, the self is always born and grows in interpersonal relationships. But no "ego" moves on into responsible selfhood if it remains chiefly the reflection of the social context around it. In our particular world in which conformity is the great destroyer of selfhood–in our society in which fitting the "pattern" tends to be accepted as the norm, and being "well liked" is the alleged ticket to salvation–what needs to be emphasized is not only the admitted fact that we are to some extent created by each other

---

6 May, *op. cit.*, p. 73.
7 *Ibid.*, p. 74.
8 *Ibid.*
9 *Ibid.*, p. 75.

but also our capacity to experience, and create, ourselves.[10]

. . . .

What does it mean to experience one's self as a self? The experience of our own identity is the basic conviction that we all start with as psychological beings. It can never be proven in a logical sense, for consciousness of one's self is the presupposition of any discussion about it...[11]

. . . .

We do not need to prove the self as an "object." It is only necessary that we show how people have the capacity for self-relatedness. The self is the organizing function within the individual and the function by means of which one human being can relate to another. It is prior to, not an object of, our science; it is presupposed in the fact that one can be a scientist.

Human experience always goes beyond our particular methods of understanding it at any given moment, and the best way to understand one's identity as a self is to look into one's own experience...[12]

. . . .

. . . the experience of one's own identity, or becoming a person, is the simplest experience in life even though at the same time the most profound...[13]

. . . .

. . .Every organism has one and only one central need in life, to fulfill its own potentialities... yet man does not grow automatically like a tree, but fulfills his potentialities only as he in his own consciousness plans and chooses.[14]

. . . .

Man, furthermore, must make his choices as an individual, for individuality is one side of one's consciousness of one's self. We can see this point clearly when we realize that consciousness of one's self is always a unique act–I can never know

---

[10] May, *op. cit.*, p. 77.
[11] *Ibid.*, p. 78.
[12] *Ibid.* p. 79.
[13] *Ibid.*, p. 80.
[14] *Ibid.*, p. 81-82.

exactly how you see yourself and you never can know exactly how I relate to myself. This is the inner sanctum where each man must stand alone. This fact makes for much of the tragedy and inescapable isolation in human life, but it also indicates again that we must find the strength in ourselves to stand in our own inner sanctum as individuals. And this fact means that, since we are not automatically merged with our fellows, we must through our own affirmation learn to love each other.[15]

. . .Consciousness of self actually expands our control of our lives, and with that expanded power comes the capacity to let ourselves go. This is the truth behind the seeming paradox, that the more consciousness of one's self one has, the more spontaneous and creative one can be at the same time.[16]

In the achieving of consciousness of one's self, most people must start back at the beginning and rediscover their feelings...[17]

Awareness of one's feelings lays the groundwork for the second step: knowing what one *wants*...[18]

The third step, along with rediscovering our feeling and wants, is to recover our relation with the subconscious aspects of ourselves...[19]

The upshot of this chapter has been to show that the more self-awareness a person has, the more alive he is. "The more consciousness," remarked Kierkegaard, "the more self." Becoming a person means this heightened awareness, this heightened experience of "I-ness," this experience that it is I, the acting one, who is the subject of what is occurring.[20]

---

[15] May, *op. cit.,* p. 82.
[16] *Ibid.,* p. 90.
[17] *Ibid.* p. 91.
[18] *Ibid.,* p. 96.
[19] *Ibid.,* p. 98.
[20] *Ibid.,* p. 100.

May then describes some of man's other unique powers. A fundamental characteristic of man is his freedom to choose. Thus, May writes :

Freedom is man's capacity to take a hand in his own development. It is our capacity to mold ourselves. Freedom is the other side of consciousness of self : if we were not able to be aware of ourselves, we would be pushed along by instinct or the automatic march of history, like bees or mastodons... Consciousness of self gives us the power to stand outside the rigid chain of stimulus and response, to pause, and by this pause to throw some weight on either side, to cast some decision about what the response will be.

That consciousness of self and freedom go together is shown in the fact that the less self-awareness a person has, the more he is unfree. That is to say, the more he is controlled by inhibitions, repressions, childhood conditionings which he has consciously "forgotten" but which still drive him unconsciously, the more he is pushed by forces over which he has no control...[21]

....

As the person gains more consciousness of self, his range of choice and his freedom proportionately increase. Freedom is cumulative; one choice made with an element of freedom makes greater freedom possible for the next choice. Each exercise of freedom enlarges the circumference of the circle of one's self.

We do not mean to imply that there are not an infinite number of deterministic influences in any one's life...But no matter how much one argues for the deterministic viewpoint, he still must grant that there is a *margin in which the alive human being can be aware of what is determining him.* And even if only in a very minute way to being with, he can have some say in how he will react to the deterministic factors.

Freedom is thus shown in how we relate to the deterministic realities of life...[22]

---

[21] May, *op. cit.,* p. 138.
[22] *Ibid.,* p. 139.

The arguments of "freedom *versus* determinism" are on a false basis, just as it is false to think of freedom as a kind of isolated electric button called "free will." Freedom is shown in according one's life with realities-realities as simple as the needs for rest and food, or as ultimate as death. Meister Eckhart expressed this approach to freedom in one of his astute psychological counsels, "When you are thwarted, it is your own attitude that is out of order." Freedom is involved when we accept the realities not by blind necessity but by choice. This means that the acceptance of limitations need not at all be a "giving up," but can and should be a constructive act of freedom; and it may well be that such a choice will have more creative results for the person than if he had not had to struggle against any limitation whatever. The man who is devoted to freedom does not waste time fighting reality; instead, as Kierkegaard remarked, he "extols reality."

. . .Freedom is most dramatically illustrated in the "heroic" actions, like Socrates' decision to drink the hemlock rather than compromise; but even more significant is the undramatic, steady day-to-day exercise of freedom on the part of any person developing toward psychological and spiritual integration in a distraught society like our own.

Thus freedom is not just the matter of saying "Yes" or "No" to a specific decision: it is the power to mold and create ourselves. Freedom is the capacity, to use Nietzsche's phrase, "to become what we truly are."[23]

The essence of existentialism, of the Sartrian as well as other varieties, is its belief in the capacity of the individual to care greatly about his freedom and inner integrity, enough to die or commit suicide for them if need be...

We agree with the fundamental Sartrian precept that the individual has no recourse from the necessity of making final decisions for himself, and that his existence as a person

---

[23] May, *op. cit.,* p. 140-142.

hangs or falls in these choices; and to make them in the last analysis in freedom and isolation may require literally as well as figuratively an agony of anxiety and inward struggle. But the fact that human beings *can* at times die for this freedom (both very strange things, quite contrary to any simple doctrine of self-preservation) implies some profound things about human nature and human existence...[24]

Freedom does not come automatically; it is achieved. And it is not gained at a single bound; it must be achieved each day... The basic step in achieving inward freedom is "choosing one's self." This strange-sounding phrase of Kierkegaard's means to affirm one's responsibility for one's self and one's existence. It is the attitude which is opposite to blind momentum or routine existence: it is an attitude of aliveness and decisiveness; it means that one recognizes that he exists in his particular spot in the universe, and he accepts the responsibility for this existence. This is what Nietzsche meant by the "will to live"–not simply the instinct for self-preservation, but the will to accept the fact that one is one's self, and to accept responsibility for fulfilling one's own destiny, which in turn implies accepting the fact that one must make his basic choices himself.[25]

When one has consciously chosen to live, two other things happen. First, his responsibility for himself takes on a new meaning. He accepts responsibility for his own life not as something with which he has been saddled, a burden forced upon him: but as a something he has chosen himself. For this person, himself, now exists as a result of a decision he himself has made. To be sure, any thinking person realizes in theory that freedom and responsibility go together: if one is not free, one is an automation and there is obviously no such thing as responsibility, and if one cannot be responsible for himself, he can't be trusted with freedom. But when one has "chosen himself," this partnership of freedom and

---

[24] May, *op. cit.,* p. 143.
[25] *Ibid.,* p. 145.

responsibility becomes more than a nice idea: he experiences it on his own pulse; in his choosing himself, he becomes aware that he has chosen personal freedom and responsibility for himself in the same breath.

The other thing which happens is that discipline from the outside is changed into *self-discipline.* He accepts discipline not because it is commanded—for who can command—and someone who has been free to take his own life?—but because he has chosen with greater freedom what he wants to do with his own life, and discipline is necessary for the sake of the values he wishes to achieve. This self-discipline can be given fancy names–Nietzsche called it "loving one's fate" and Spinoza spoke of obedience to the laws of life. But whether bedecked by fancy terms or not, it is, I believe, a lesson everyone progressively learns in his struggle toward maturity.[26]

The next characteristic that May mentions is man's conscience. Thus, he notes :

Man is the "ethical animal"–ethical in potentiality even if, unfortunately, not in actuality. His capacity for ethical judgement—like freedom, reason and the other unique characteristics of the human being–is based upon his consciousness of himself.[27]

Man can "look before and after." He can transcend the immediate moment, can remember the past and plan for the future, and thus choose a good which is greater, but will not occur till some future moment in preference to a lesser, immediate one. By the same token he can feel himself into someone else's needs and desires, can imagine himself in the other's place, and so make his choices with a view to the good of his fellows as well as himself. This is the beginning of the capacity, however imperfect and rudimentary it may be in most people, to "love thy neighbor" and to be aware of the relation between their own acts and the welfare of the community.

The human being not only *can* make such choices of values and goals, but he is the animal who *must* do so if he is to attain

---

[26] May, *op. cit.,* pp. 148-149.
[27] *Ibid.,* p. 150.

integration. For the value–the goal he moves toward–serves him as a psychological center, a kind of core of integration which draws together his powers as the core of a magnet draws the magnet's lines of force together. We pointed out in a previous chapter that knowing what one *wants* is essential for the beginnings of the child's and young person's capacity for self-direction. Knowing what one wants is simply the elemental form of what in the maturing person is the ability to choose one's own values. The mark of the mature man is that his living is integrated around self-chosen goals: he knows what he wants, no longer simply as the child wants ice cream but as the grown person plans and works toward a creative love relationship or toward business achievement or what not. He loves the members of his family not because he has been thrown together with them by the accident of birth but because he finds them lovable and chooses to love them; and he works not merely from automatic routine but because he consciously believes in the value of what he is doing.[28]

Man is the ethical animal: but his achievement of ethical awareness is not easy. He does not grow into ethical judgement as simply as the flower grows toward the sun. Indeed, like freedom and the other aspects of man's consciousness of self, ethical awareness is gained only at the price of inner conflict and anxiety.[29]

Conscience is not a set of handed-down prohibitions to constrict the self, to stifle its vitality and impulses. Nor is conscience to be thought of as divorced from tradition, as in the liberalistic period when it was implied that one decided every act *de novo*. Conscience, rather, is one's capacity to tap one's own deeper levels of insight, ethical sensitivity and awareness, in which tradition and immediate experience are not opposed to each other but interrelated. The etymology of the term reveals this point. Composed of the two Latin words

---

[28] May, *op. cit.,* p. 151.
[29] *Ibid.,* p. 155.

meaning " to know" (scire) and "with" (cum), conscience is very close to the term consciousness. . .

We wish thus to emphasize the positive aspects of conscience—conscience as the individual's method of tapping wisdom and insight within himself, conscience as an "opening up," a guide to enlarge experience. This is what Nietzsche was referring to in his paean on the theme "*beyond* good and evil," and what Tillich means in his concept of the *transmoral* conscience. With this view it will no longer be true that "conscience doth make cowards of us all." Conscience, rather, will be the taproot of courage.[30]

May proceeds to discuss the quality of courage :

In any age courage is the simple virtue needed for a human being to traverse the rocky road from infancy to maturity of personality. But in an age of anxiety, an age of herd morality and personal isolation, courage is a *sine qua non.*[31]

Courage is the capacity to meet the anxiety which arises as one achieves freedom. It is the willingness to differentiate, to move from the protecting realms of parental dependence to new levels of freedom and integration. The need for courage arises not only at those stages when breaks with parental protection are most obvious–such as at the birth of self-awareness, at going off to school, at adolescence, in crises of love, marriage and the facing of ultimate death–but at every step in between as one moves from the familiar surroundings over frontiers into the unfamiliar. "Courage, in its final analysis," as the neurobiologist Dr. Kurt Goldstein well puts it, "is nothing but an affirmative answer to the shocks of existence which must be borne for the actualization of one's own nature."

The opposite to courage is not cowardice: that, rather, is the lack of courage. To say a person is a coward has no more meaning than to say he is lazy: it simply tells us that some vital potentiality is unrealized or blocked. The opposite to

---

[30] May, op. cit., p. 184.
[31] *Ibid.,* p. 191.

courage, as one endeavors to understand the problem in our particular age, is automaton conformity.

The courage to be one's self is scarcely admired as the top virtue these days...[32]

....

What we lack in our day is an understanding of the friendly, warm, personal, original, constructive courage of a Socrates or a Spinoza. We need to recover an understanding of the positive aspects of courage—courage as the inner side of growth, courage as a constructive way of that becoming of one's self which is prior to the power to give one's self. Thus, when in this chapter we emphasize standing on one's own belief, we do not at all imply living in a vacuum of separateness; actually, courage is the basis of any creative relationship...[33]

....

Courage is necessary in every step in a person's movement from the mass–symbolically the womb–to becoming a person in his own right; it is at each step as though one suffers the pangs of his own birth. Courage, whether the soldier's courage in risking death or the child's in going off to school, means the power to *let go* of the familiar and the secure. Courage is required not only in a person's occasional crucial decision for his own freedom, but in the little hour-to-hour decisions which place the bricks in the structure of his building of himself into a person who acts with freedom and responsibility.[34]

....

. . .it requires greater courage to preserve inner freedom, to move on in one's inward journey into new realms, than to stand defiantly for outer freedom. It is often easier to play the martyr, as it is to be rash in battle. Strange as it sounds, steady, patient growth in freedom is probably the most difficult task of all, requiring the greatest courage..[35]

May then moves on to appreciate man's ability to live in the

---

[32] May, *op. cit.*, p. 192.

[33] *Ibid.*, p. 193

[34] *Ibid.*, p. 196.

[35] *Ibid.*, p. 197.

present, yet be able simultaneously to imagine the future and utilize the past. He notes ;

The power to "look before and after" is part of man's ability to be conscious of himself. Plants and animals live by quantitative time: an hour, a week or a year past, and the tree has another ring on its trunk. But time is a quite different thing for human beings; man is the time surmounting mammal. In his work on semantics, Alfred Korzybski has insistently made the point that the characteristic which distinguishes man from all other living things is his *time-binding* capacity. By that, says Korzybski, "I mean the capacity to use the fruits of past labors and experiences as intellectual or spiritual capital for developments in the present... I mean the capacity of human beings to conduct their lives in the ever increasing light of inherited wisdom; I mean the capacity in virtue of which man is at once the heritor of the by-gone ages and the trustee of posterity."[36]

Psychologically and spiritually, man does not live by the clock alone. His time, rather, depends on the *significance* of the event... Psychological time is not the sheer passage of time as such, but the *meaning* of the experience, that is, what is significant for the person's hopes, anxiety, growth.[37]

....

. . .The less alive a person is— "alive" here defined as having conscious direction of his life—the more is time for him the time of the clock. The more alive he is, the more he lives by qualitative time.[38]

May concludes by mentioning that :

The first thing necessary for a constructive dealing with time is to learn to live in the reality of the present moment. For psychologically speaking, this present moment is all we have. . .[39]

When a person looks directly into himself, all he is aware of is his instant of consciousness at that particular moment of the present.

---

[36] May, *op. cit.,* p. 219.
[37] *Ibid.,* p. 219-220.
[38] *Ibid.,* p. 221.
[39] *Ibid.,* p. 226.

It is this instant of consciousness which is most real, and must not be fled from.[40]

. . . .

It is by no means as easy as it may look to live in the immediate present. For it requires a high degree of awareness of one's self as an experiencing "I". The less one is conscious of himself as the one who acts, that is, the more unfree and automatic he is, the less he will be aware of the immediate present. . .

But the more awareness one has—that is, the more he experiences himself as the acting, directing agent in what he is doing—the more alive he will be and the more responsive to the present moment. Like self-awareness itself, this experiencing of the quality of the present can be cultivated. . .[41]

. . . .

The most effective way to ensure the value of the future, as we have mentioned, is to confront the present courageously and constructively... an essential characteristic of the creative act done in human consciousness is that it is not limited by quantitative time. No one values a painting according to how long it took to paint it or how big it is: should we judge our actions by more superficial standards than a painting?[42]

. . . .

The present moment is thus not limited from one point on the clock to another. It is always "pregnant," always ready to open, to give birth. One has only to try the experiment of looking deeply within himself, let us say, trailing almost any random ideas, and he will find, so rich is a moment of consciousness in the human mind, that associations and new ideas beckon in every direction... the moment always has its "finite" side, to use a philosophical term, which the mature person never forgets. But the moment also always has its infinite side, it always beckons with new possibilities. Time for the human being is not a corridor; it is a continual opening out.[43]

---

[40] May, *op. cit.*, p. 227.
[41] *Ibid.*, pp. 227-228.
[42] *Ibid.*, p. 229.
[43] *Ibid.*, p. 230.

TABLE 1 MAY'S VIEW OF MAN : CONTENT ANALYSIS

No.	pp. 114-115	pp. 116-118	pp. 119-122	pp. 123-124	p. 125	pp. 126-127	Total
1.		1					1
2.	3						3
3.							0
4.		8	2	1		3	14
5.		1					1
6.							0
7.							0
8.							0
9.							0
10.							0
11.							0
12.							0
13.							0
14.			1	6			7
15.							0
16.		1					1
17.							0
18.						5	5
19.							0
20.					9		9
21.							0
22.		1	1				2
23.		1					1
24.		2	7	1			10
25			1				1
26.							0
27.			1	6			7
28.							0
29.		2					2
30a.			1				1
30b.		1					1
					Cumulative total		66

Note: The concept of 'conscience' has been coded as: ontologically and socially responsible.

TABLE 2 MAY'S VIEW OF MAN: PERCENTAGE EMPHASIS

No.	Dimension/Category	Total	Percentage Degree of Emphasis
1.	A communicative being	1	1.5
2.	Accepting of non-hedonic emotions	3	4.5
3.	Always in process	0	0
4.	Conscious/Aware	14	21.2
5.	Creative	1	1.5
6.	Forward-thrusting	0	0
7.	Freedom-cherishing	0	0
8.	Future-imagining	0	0
9.	Goal-directed	0	0
10.	Holistic/An integrated whole	0	0
11.	Intelligent	0	0
12.	Much like fellow man	0	0
13.	Nature-appreciating	0	0
14.	Ontologically responsible	7	10.5
15.	Open to life/experience	0	0
16.	Other-affirming	1	1.5
17.	Pleasure-loving	0	0
18.	Present-confronting	5	7.5
19.	Rational	0	0
20.	Risk-taking/Courageous	9	13.5
21.	Self-active	0	0
22.	Self-actualizing/Self-realizing	2	3.0
23.	Self-affirming	1	1.5
24.	Self-determining	10	15.0
25.	Self-disciplining	1	1.5
26.	Sexual	0	0
27.	Socially responsible	7	10.5
28.	Ultimately unknowable	0	0
29.	Unique	2	3.0
30a.	Reality-accepting	1	1.5
30b.	Ultimately alone	1	1.5
		66	99.2*

*This amount differs from 100 percent because the figures were rounded off to the first decimal place for the sake of clarity.

SUMMARY

This chapter has consisted of the summarized version of May's view of man and its analysis by means of the specially designed content analysis instrument.

In his view of man, May has given primary emphasis (10 percent or more emphasis) to the following dimensions:

Man is conscious/aware

Man is self-determining

Man is risk-taking/courageous

Man is ontologically responsible

Man is socially responsible

Secondary emphasis (between 5 and 10 percent emphasis) has been given to the dimension:

Man is present-confronting

The remaining dimensions received tertiary emphasis (less than 5 percent emphasis).

It appears that May has given only tertiary emphasis to the concept that man is self-actualizing. This would, however, be an erroneous interpretation of his view of man for he has specifically stated that:

. . .Every organism has one and only one central need in life, to fulfill its own potentialities. . .[44]

May sees full self-actualization, in the sense of fully utilizing his potentialities, as the primary need-value-task of man. In his view of man, May has given primary emphasis to those powers of man which he must particularly use in order to effectively satisfy his one central need.

Like Bugental, May considers man's consciousness to be his most distinctive source of guidance for the choices that he must make to satisfy his hunger for self-actualization. In addition to this, however, he places an equal degree of emphasis on man's conscience — his feeling of responsibility for facilitating both his

---

[44] May, *op. cit.*, p. 81.

own and his fellow man's self-actualization. Finally, he gives primary emphasis to the part that courage plays in shaping the manner in which man meets the anxieties of existence and fulfills his potentialities. Needless to say, the powers of consciousness, conscience and courage are, like all concepts concerned with man, deeply interrelated and interdependent. The use of any one of them fosters the use of the others, and simultaneously, the growth and fulfillment of the person as a whole.

9

SYNTHESIS

The synthesis has been constructed by first developing two comparison tables depicting the areas and strength of agreement and disagreement between the five views of man presented in the preceding chapters.

In order to develop the first comparison table, two groups of dimensions have been used:

1. the dimensions to which a scientist has given 10 percent or more emphasis;
2. the dimensions to which he has given between 5 and 10 percent emphasis.

The former dimensions are also called areas of primary emphasis, while the latter are called areas of secondary emphasis. Each of the five views of man will now be individually presented. These will be followed by the two comparison tables; finally, the synthesis will be formulated.

HUBERT BONNER'S VIEW OF MAN

In Bonner's view, the areas of primary emphasis are:

Man is self-determining	:	18.9%
Man is self-actualizing	:	15.6%
The areas of secondary emphasis are	:	
Man is always in process	:	7.8%
Man is unique	:	6.7%
Man is forward-thrusting	:	5.6%
Man is socially responsible	:	5.6%

JAMES F.T. BUGENTAL'S VIEW OF MAN

In Bugental's view, the areas of primary emphasis are:

Man is self-determining	:	16.3%
Man is conscious/aware	:	15.2%

The areas of secondary emphasis are   :
Man is accepting of non-hedonic emotions   :   9.8%
Man is a communicative being   :   5.4%
Man is self-actualizing   :   5.4%

ERICH FROMM'S VIEW OF MAN

In Fromm's view, the areas of primary emphasis are:
Man is self-actualizing   :   24.3%
Man is rational   :   12.6%
Man is self-affirming   :   10.8%
The areas of secondary emphasis are   :
Man is conscious/aware   :   8.1%
Man is self-determining   :   7.2%

ABRAHAM H. MASLOW'S VIEW OF MAN

In Maslow's view, the areas of primary emphasis are:
Man is self-actualizing   :   16.8%
Man is self-affirming   :   14.4%
Man is holistic/An integrated whole   :   12.0%
Man is other-affirming   :   12.0%
Man is unique   :   12.0%
None of the dimensions were given between 5 percent and 10 percent emphasis.

ROLLO MAY'S VIEW OF MAN

In May's view, the areas of primary emphasis are :
Man is conscious/aware   :   21.2%
Man is self-determining   :   15.0%
Man is risk-taking/courageous   :   13.5%
Man is ontologically responsible   :   10.5%
Man is socially responsible   :   10.5%
The following dimension received secondary emphasis :
Man is present-confronting   :   7.5%

The index to the first comparison table which appears in Table 1 is given below :

Group A consists of areas that have received primary emphasis from three scientists (*i.e.,* three primary mentions) and secondary emphasis from one scientist (*i.e.,* one secondary mention).

Table 1 COMPARISON OF THE FIVE VIEWS OF MAN USING AREAS/DIMENSIONS THAT RECEIVED A MINIMUM OF ONE SECONDARY MENTION

Areas/Dimensions	Bonner	Bugental	Fromm	Maslow	May
Group A					
Man is self-actualizing	15.9	5.4	24.3	16.8	3.0
Man is self-determining	18.9	16.3	7.2	4.8	15.0
Group B					
Man is conscious/aware	2.2	15.2	8.1	0	21.2
Group C					
Man is self-affirming	4.4	3.3	10.8	14.4	1.5
Group D					
Man is unique	6.7	1.1	3.6	12.0	3.0
Man is socially responsible	5.6	4.3	0	4.8	10.5
Group E					
Man is risk-taking/courageous	2.2	4.3	0	0	13.5
Man is rational	0	0	12.6	0	0
Man is holistic/An integrated whole	4.4	4.3	0.9	12.0	0
Man is other-affirming	4.4	2.2	3.6	12.0	1.5
Man is ontologically responsible	3.3	4.3	2.7	2.4	10.5
Group F					
Man is always in process	7.8	4.3	0.9	0	0
Man is forward-thrusing	5.6	0	0	0	0
Man is accepting of non-hedonic emotions	2.2	9.8	0	2.4	4.5
Man is communicative being	0	5.4	4.5	0	1.5
Man is present-confronting	0	3.3	0	0	7.5

NOTE : All the figures in this table are percentages.

Group B consists of areas given two primary mentions and one secondary.

Group C consists of areas given two primary mentions.

Group D consists of areas given one primary mention and one secondary.

Group E consists of areas given one primary mention.

Group F consists of areas given one secondary mention.

Some of the more significant points that the first comparison table highlights are:

1. There is very strong emphasis that man is self-actualizing and self-determining. (See Group A).

2. There is strong emphasis that man is conscious/aware, self-affirming, unique and socially responsible. (See Groups B, C and D)

3. In the case of each of the areas in Group E, one scientist has seen fit to give it primary emphasis in his view of man. The areas are listed below:

Man is risk-taking/courageous

Man is rational

Man is holistic/An integrated whole

Man is other-affirming

Man is ontologically responsible

And, each of the latter three dimensions have been mentioned by at least three other scientists.

4. In the case of each of the areas in Group F, one scientist has given it secondary emphasis in his view of man. The areas are :

Man is always in process

Man is forward-thrusting

Man is accepting of non-hedonic emotions

Man is a communicative being

Man is present-confronting

And, in the case of the dimension 'accepting of non-hedonic emotions' three other scientists have also mentioned it.

5. There are a total of 15 dimensions that have not been included in the above six groups. A second comparison table has been developed to portray these dimensions

each of which had received tertiary emphasis (less than 5 percent emphasis) by the scientists in their respective views.

TABLE 2 COMPARISON OF THE FIVE VIEWS OF MAN USING AREAS/DIMENSIONS THAT RECEIVED TERTIARY EMPHASIS

Areas/Dimensions	Bonner	Bugental	Fromm	Maslow	May
Creative	2.2	2.2	2.7	2.4	1.5
Freedom-cherishing	1.1	0	0.9	0	0
Future-imagining	2.2	0	0.9	0	0
Goal-directed	4.4	2.2	0.9	4.8	0
Intelligent	0	1.1	2.7	0	0
Much like fellow man	0	2.2	1.8	2.4	0
Nature-appreciating	0	0	0	2.4	0
Open to life/experience	2.2	1.1	0	0	0
Pleasure-loving	0	1.1	0	0	0
Self-active	0	0	0.9	2.4	0
Self-disciplining	3.3	1.1	1.8	0	1.5
Sexual	0	0	0	0	0
Ultimately unknowable	1.1	0	0	0	0
Reality-accepting	0	3.3	3.6	4.8	1.5
Ultimately alone	0	2.2	4.5	0	1.5

NOTE : All the figures in this table are percentages.

From the second comparison table we distinguish the following two groups of dimensions:

Group G (dimensions that have been mentioned by at least four scientists).

Man is creative
Man is goal-directed
Man is self-disciplining
Man is reality-accepting

Group H (dimensions that have been mentioned by less than four scientists)

Man is much like fellow man
Man is ultimately alone
Man is freedom-cherishing

Man is future-imagining
Man is intelligent
Man is nature-appreciating
Man is open to life/experience
Man is pleasure-loving
Man is self-active
Man is ultimately unknowable
Man is sexual

On the basis of the two comparison tables that have been presented above, the synthesis will now be formulated. One basic criterion needs to be noted:

A dimension will be included in the synthesis only if it has received mention by at least four of the five scientists.

Thus, the synthesis encouraged by the foregoing analysis of the views of man held by a group of five leading, contemporary, humanistically-oriented, behavioral scientists, is as follows:

1. There is very strong emphasis that: man is self-actualizing and self-determining.
2. There is strong emphasis that: man is conscious/aware, self-affirming, unique and socially responsible.
3. There is some degree of emphasis that: man is holistic/an integrated whole, other-affirming, ontologically responsible, and accepting of non-hedonic emotions.
4. There is a low degree of emphasis that: man is creative, goal-directed, self-disciplining, and reality-accepting.

SUMMARY

This chapter has consisted of:
1. the comparison of the areas and strength of agreement and disagreement between the five views of man; and,
2. the formulation of a synthesis on the basis of the above mentioned comparative analysis of the five theoretical positions.

10

J. BRONOWSKI'S COMMENT

This chapter consists of an evaluation of the synthesis and dissertation by J. Bronowski. In addition to this, an unpublished paper of his entitled, "A Twentieth Century Image of Man" has been included as an example of a natural scientist's view of man.

COMMENT ON THE DISSERTATION

The dissertation is an original and searching attempt to form a rounded image of man as pictured by contemporary humanists. The author uses a rigorous, statistical method which is itself an innovation in his field, on which he is to be congratulated. It is clear that analysis by this method is a powerful tool, which can with advantage be applied widely to the work of groups of thinkers; and I have no doubt that the method will now be followed by other students.

Turning to the conclusions, a natural scientist is struck immediately by the homogeneity of the preoccupations of the five writers whose work has been examined here. The results show far less scatter than I should have expected, both in what is stressed and what is neglected. What is stressed by all five writers, it turns out, is consistently the human search for fulfillment *within the self*. What is neglected is all that range of satisfactions which man finds in his *exploration of nature*. Behavioral scientists (judging from this distinguished sample) are indifferent to man's search to understand nature, and particularly to the *means by which man relates to external nature*.

For example, it is astonishing to a natural scientist that there should be no stress on knowledge as a means of human fulfillment. Only one of the five authors puts rational thought in his area of primary emphasis, and none of them puts intelligence in it. Only one of the authors puts human communication into any area of emphasis at all, and none discusses the relation between knowledge and language—either the outer language of public communication or the inner language of discovery. And, most remarkably, only one puts the appreciation of nature even into the area of tertiary interest! It is evident that natural scientists with a humanistic outlook have an entirely different emphasis: Albert Einstein, for example, Bertrand Russell, or Julian Huxley. Nor is this difference peculiar to modern scientists; three hundred years ago Leibnitz was already defining man in terms which stress his sympathy with the laws of nature, and the philosophy of Spinoza is grounded in the same sympathy.

In one respect, however, the difference between the behavioral and the natural scientist seems to be wider now. The five writers examined seem (to a contemporary natural scientist) to be out of touch with changes in man's self image that have taken place as a result of new findings in anthropology, and particularly in the evolution of man, in the last thirty years. For example, sex as a human preoccupation gets not a single vote even at the tertiary level from these authors; yet sex in the human sense (the prohibition of incest, love, monogamy, intellectual sympathy, family organization and education) has evidently been a major selective force in directing human evolution towards cultural instead of biological change. This is one aspect of what a natural scientist today sees as the future-directed orientation of man; yet of the five behavioral scientists, only one gives this secondary emphasis, and two by contrast reserve their emphasis for man as "present-confronting."

In this context, I ought to make some reference to what has been learnt in recent years from animal studies. This is a disputed field at present, because those who have popularized it have paid too little attention to the crucial differences in behavior between different animal species. Nevertheless, to a natural scientist, it is unnatural to characterize man without referring to what he shares with other animals, and what makes him different. For example, there is evidence of communication between members in every species; but only in man is this outer language also transformed into an inner language for self-communication and exploration. Yet the five behavioral scientists show no interest in such questions, which natural scientists believe to be the most promising source for the future study of man's special gifts.

In summary, I have found this dissertation fascinating as a work of research and scholarship, and deeply disturbing as an account of the preoccupation of leading behavioral scientists who are humanists. It is plain that the gap between what is glibly called "The Two Cultures" is by no means closed within the sciences. Certainly natural scientists have much to learn from behavioral scientists—including a becoming modesty; but just as certainly behavioral scientists are, on this evidence, preoccupied with aspects of the human mind which ignore rational knowledge and imaginative exploration of nature. As a natural scientist, I find that distressing, and as a humanist, it seems to me a surrender to the irrational.

signed J. Bronowski

15 September 1972

P.S. It happens that I have recently written an essay on "A Twentieth Century Image of Man," and it may be useful if I append it to these notes as an example of a natural scientist's view.

A TWENTIETH CENTURY IMAGE OF MAN

History does not count in centuries, and sets no more store by the number twenty then by any smaller number. It follows that such a handy descriptive phrase as "twentieth century man," for all its bold ring, is not self-sufficient, or even self-explanatory. The culture of the twentieth century that we now form and transform and reform is not easy to characterize; it is a fascinating and strenuous interplay of the old and the new. For example, one way to describe our civilization is to point to the high content of scientific knowledge in it, and the wide spread of education that goes with that—so that we live for the first time in a democracy of knowledge. And yet the democracy of knowledge did not begin in the twentieth century: it began 500 years ago, with the spread of the printed book after 1450.

Certainly the power of natural, empirical knowledge is more massive in our society than ever in the past. Yet it has come by steps that can be identified and dated, and they are of a very respectable age. We can date the technical innovations, of course, all the way from the flying shuttle and the steam engine to nuclear energy. And we can also date the intellectual components that have gone to make up the scientific outlook of the century: the publication by Isaac Newton of the *Principia* in

1687, for example, and by Charles Darwin of the *Origin of Species* in 1859. If we really want to grasp what is new in our world, and to ask how it should affect our image of man, we must push the analysis deeper.

The central principle that moves our society, and has done so since the decline of scholastic authority at the Reformation, is the demand that the basis of action shall be verifiable knowledge. The strong word here, the post-Renaissance word, is verifiable. In the Middle Ages, and before, a great deal of knowledge was accepted on authority: this is evident in matters of religion, of course, and in abstract subjects like astronomy, but it is just as remarkable in a practical science like medicine, which in the year 1500 was still taught from books written before the year 200. When Paracelsus publicly burned these medical classics in Basel in 1527, he was setting up the sanction of verifiable knowledge.

To verify knowledge cannot mean to prove it absolutely true: it is not given to human beings to do that. Our powers are confined to a more modest endeavor, which is to check the assertions of others in our own experience. In this way we turn private conviction into public knowledge, and create a democracy of knowledge; the demand that knowledge shall be verifiable is simply a demand that the basis for public action shall be accessible to the independent scrutiny of every member of the public. Knowledge which is open to public inspection in this way has come to be called science, but that Latin name does not change its character as the mode of knowledge that is accessible to all. There is no other way to establish a democracy of knowledge and, conversely, the free spread of such knowledge is a condition for democracy.

Since I am concerned with mental images as well as with actions, I must not pass over in silence the desolate and damaging image of science that the demand for verification has conjured up. To test what is conjectured or asserted requires a painstaking apparatus: a stop-watch, a set of measures, a slide-rule, all the formidable array of glassware and reagents and electrical devices that ensures that a laboratory will be impersonal. These mechanical watchdogs against bias have nothing to do with the content of the knowledge that we are at pains to share; they merely represent and in some way symbolize the procedures by which knowledge becomes verifiable, and thereby open to the public. The irony and perhaps the tragedy of science is that, as symbols, the test-tubes and the balances of the laboratory have quite the opposite effect: by insisting that knowledge, to be shared, must be stated precisely, they make science look forbidding and worse, esoteric.

The preoccupation with the mechanics of the laboratory has naturally brought in its train a mechanical picture of the activity that is science and, by implication, of scientific knowledge. By the end of the nineteenth century, the man in the laboratory coat had come to be regarded as a living extension of his own apparatus. Working mainly in the physical sciences, he seemed to be reducing nature to a machine; and it was expected that in time he would prove that man is a machine also would in fact turn man into a machine, as a kind of projection of his own glacial temperament.

I have remarked that this conventional picture of the scientist, and of man in his image, was formed in the nineteenth century; the classical expressions in literature of the pair of them are, I suppose, Frankenstein by Mary Shelley, and Phineas Fogg by Jules Verne. Even the younger of these is already a hundred years old (Verne published *Around the World in Eighty Days* as a serial in 1872), yet the model of man as a machine has remained dominant, and goes on haunting the imagination of most people to this day. Evidently the image of man as a machine has the special psychological power of ambivalence; it fascinates most those whom it most repels. There is no other way to explain its perverse popularity, all the way from science fiction to the writings on animal behavior by Konrad Lorenz and on human behavior by B.F. Skinner.

The effect of this strange and irrational, one might almost say masochistic, persistence of a nineteenth century image has been catastrophic. Nothing else seems to me to have contributed so much to the erosion of self-confidence and the loss of nerve with which our civilization is now struggling. The liberal arts have been impoverished and drained of talent by what is in effect an ideological retreat from the most lively areas of modern knowledge. The young have joined their reactionary elders in a facile abuse of all technical innovation, without asking themselves how it has ever come about that their assertion of self (which was unknown a hundred years ago) is now heard and heeded. The shift in education that has made this possible, the principle that what should be taught is not vocation but knowledge, not an ability to do but to think, is now derided as irrelevant. The democratic belief that every man is entitled to the clean air and the green landscape that the rich have always guarded for themselves is turned into a blanket crusade against all technology, although technology has been the main instrument that has made democracy possible — that has made democracy viable, intellectually as well as materially. The intellectual step from technology to science, which set our civilization in motion after the Middle Ages, is minimized even in works of scholarship, such as those of Lynn Thorndike on magic and Joseph Needham on China — and it is hardly surprising that those cultures from the past are then admired by anti-intellectuals. Most deeply and most sadly, the foundations for a liberal ethic are still sought in the past, and thereby separated from the living sources that the new knowledge of the twentieth century can create for them.

My strictures imply that the model of science that has been handed down from the last century is false, and must be replaced. Not only has science made new discoveries, and opened new fields of knowledge: that was to be expected. What was unexpected is the new realization that public knowledge, verifiable knowledge, does not and cannot have the inhuman, mechanical kind of exactness that was ascribed to it in the nineteenth century.

The realization that scientific knowledge must of necessity have a personal component came first in physics; and although physics is not central in my theme, I ought to glance at it. When quantum physics first made the headlines, early in this century, what was stressed was the breach that the new work opened in the classical doctrine of determinism; its expositors liked to suggest that through this loop-hole free will might slip into the scientific universe. But that vague speculation, even if it means anything (which I doubt), is beside the point. Werner Heisenberg in 1927 put his finger on the intellectual point in quantum mechanics: the principle that no description of an event, however minute, can be complete. Our knowledge of the real world is verifiable, yes and we can exchange information about what we see in the world, but only with a finite tolerance. There is an insurmountable limit to the precision with which we can formulate experience and make it public.

In a deep sense, this had already been said by Albert Einstein when he announced the principle of relativity in 1905. Again, I do not want to labor a concept in physics, except to underline its human significance. Einstein showed that the laws of physics are universal, that is, are formulated in the same terms by every observer—but only because he carries his own universe with him. Time as you measure it may be different from my time, mass as you measure it may be different from my mass, speed and momentum and energy may all be different; it is only the relations between them that remain the same for us both. Each of us rides his personal universe, his own travelling box of time and space, and all that they have in common is the same structure or coherence; when we formalize our experiences, they yield the same laws.

These principles express a far-reaching revision in the idea of knowledge. They shift the emphasis away from the impersonal record, and they put in its place a relation from which the human observer cannot be abstracted. The scrutiny of experience is no longer idealized as an activity that could be carried out by a machine. There is no reality, there are no laws, that can be separated from the process of their discovery; the human condition is also the necessary condition for the recognition of order in the world.

The place of man in the world has become a subject of driving scientific interest for another reason. As the century has gone forward, the preoccupation of science has been moving from physics to biology. The watershed came after 1945, when many young scientists turned away from their war work and seized the chance to enter a new and different kind of field. About this time, it was found that the basic carrier of heredity in all animals is not (as had been supposed) a complex protein, but a fairly simple nucleic acid. When the geometrical structure of the acid was unravelled by Francis Crick and James Watson shortly after, in 1953, biology came of age as an intellectual discipline; and it has attracted the best minds ever since.

We naturally study first the machinery of life that is shared by all species: for example, we have learned much in the last twenty years about the cellular biology of man by studying some bacteria. But it is also important and revealing to ask what in his biological equipment has made man take a different path from the other animals—has made him, for example, adapt his environments to himself instead of the other way about. Man is the newcomer in biological evolution; his first recognizable fossil ancestor (found by Raymond Dart in Africa in 1924) is not much more than two million years old. How have we come so far in so short a time? Questions like this have transformed the image of man in the twentieth century. It has become clear that man is unique in being predominantly a cultural animal. Even his biological evolution has been culture-driven: by which I mean that evolution has taken the direction to *homo sapiens* as a result of the elaboration of a cultural skill that his early ancestors discovered, namely the manual dexterity to make stone tools and the mental foresight and imagery to store them for use in the future. The enlargement of the brain, the development of speech, the march over the whole globe, the use of fire, and the planning first of hunting and then of agriculture, all flow from that beginning. More recently man has been able to make cultural evolution take the place of biological evolution altogether — since the end of the last Ice Age, roughly ten thousand years ago.

A central component in this progress, this ascent of man, has been the manipulation of mental imagery and its expression in language. There is a biological locus for that in and around the speech areas in the brain, which are unique to man; and the projection into the future which they serve is localized in another adjacent part of the human brain, his great frontal lobes. These structures testify that *homo sapiens* is rightly distinguished as a planning animal, and that his plans depend on a rational analysis of the world of the kind that is formalized in his languages — including the languages of mathematics and science. In this sense, man is unique because he prepares for his actions by seeking knowledge, and is able to separate that from the other responses which his environment evokes.

I come to rest on these two concepts: planning and knowledge. Since we now understand that human knowledge is necessarily incomplete, it follows that our plans are not simply calculations. A calculation is, as it were, a tactical plan, to solve an immediate and finite problem of action. But the large problems of conduct which shape our lives are not immediate and finite; and for them we have to devise much more general plans, namely the grand strategies that we call values. This is how we derive and this is what we mean by such values as justice, loyalty, devotion, respect, dignity, affection, and integrity. Values are the strategies by which we guide our conduct in the face of the insoluble problems of human relations, and walk the knife-edge between our solitary desires and our social needs. Thus values are now seen to be as integral a part of human nature, the biological nature of man, as is a large brain or our inability to oxidize uric acid in our cells.

There is a familiar argument in philosophy, usually ascribed to G.E. Moore, which asserts that values cannot be derived from knowledge: to argue from "what is" to "how we ought to behave" is stigmatized as a naturalistic fallacy. What I have constructed in this twentieth century image of man is a different and lesser argument, namely that we now understand that the elaboration of values is as characteristic a human activity as is the search for knowledge. But we can take the argument further, and show that it is a philosophical error and a misreading of the nature of reality to suppose that values are independent of knowledge. The fact, the scientific fact is that knowledge does not exist until we search for it, and the phrase "what is" has no meaning for us until we take pains to discover it. We learn to know, we learn what is, only by behaving in a certain way: and how we ought to behave is therefore prescribed in a fundamental way by our search to know what is. Human beings create their values, they form an ethic, because they direct their aspirations towards the command of nature, not by the means which other animals use, but by the road of knowledge. And that road is a biological necessity for us, it is our world line, which characterizes our nature as human nature. The image of man has become clear and exhilarating in the twentieth century, and I will describe it in a picture that I have drawn once before.

"Like the other primates, man is noisy, inquisitive, cooperative, intelligent, skillful, thoughtful, and as busy with himself as with his environment. These features are not common in the rest of the animal world, singly or in combination. They have been a great deal more important in the evolution of the primates than the territorial imperative and the aggressive drives which we share with lower animals. And in the remarkable order of primates, the evolution of man is most remarkable and spectacular. His gifts of discrimination and judgment, the ability to speak, to remember, to foresee, to imagine and to think symbolically, his carriage and the freedom that it gives to hands and face, his face-to-face relations and his way of making love, his family life and the intimacy of his social values, are an incomparable biological equipment. They have evolved him, and in turn have been evolved by his own progress, within at most a few million years. From them man has his creative skill and his imaginative breadth of outlook, in which are intertwined his need for the society of others and his urge to think for himself.

"What makes the biological machinery of man so powerful is that it modifies his actions through his imagination: it makes him able to symbolize, to project himself into the consequences of his acts, to conceptualize his plans, and to weigh them one against another as a system of values. We are the creatures who have to create values in order to elucidate our own conduct, and to learn from it, so that we can direct it into the future."

SUMMARY

This chapter has consisted of J. Bronowski's comment on the synthesis and an article by him entitled, "A Twentieth Century Image of Man" portraying a natural scientist's view of man.

In his comment, Bronowski emphasizes that whereas all five behavioral scientists have stressed man's desire to become more of what he authentically and potentially already is, they have neglected to stress man's desire to explore and communicate with external nature, his desire to use his reason, knowledge and intelligence to appreciatively understand and constructively utilize the laws of nature. Such overemphasis on phenomena within the self and underemphasis of phenomena outside the self distresses Bronowski and he concludes with the words, ". . .it seems to me a surrender to the irrational."

And in the article, Bronowski emphasizes that man, the predominantly culturally evolved animal, is distinguished by his planning and knowledge-seeking nature. Through his ability to plan man has, over the past few million years, incorporated values like affection, respect, justice, etc., into his biological nature as surely as he has created for himself a large brain.

11

SUMMARY AND DISCUSSION

This study was designed to obtain a better understanding of the view of man held by leading, contemporary, humanistic behavioral scientists.

The problem posed by the study was stated in the form of three questions:

I. What can be determined regarding the areas and strength of agreement and disagreement in the views of man held by a group of leading, contemporary, humanistically-oriented, behavioral scientists?

II. What is the synthesis obtained from an analysis of a selection of contemporary, humanistic views of man?

III. Given a synthesis of the humanistic view of man what would be the evaluation of a natural scientist who is a recognized authority?

The research methodology used for answering these questions was to design an appropriate instrument for content analysis and then to use this on summarized versions of the views of man expressed by the five selected humanistic psychologists. This enabled a comparative evaluation of their views of man; and, on the basis of this analysis, the synthesis was formulated. This synthesis was then presented to J. Bronowski for a critical appraisal.

The findings with respect to the first question have been catalogued in the two comparison tables in chapter 9.

The findings related to the second question are as follows:

1. There is very strong emphasis that: man is self-actualizing and self-determining.

2. There is strong emphasis that: man is conscious/aware, self-affirming, unique, and socially responsible.

3. There is some degree of emphasis that: man is holistic/an integrated whole, other-affirming, ontologically responsible, and accepting of non-hedonic emotions.

4. There is a low degree of emphasis that: man is creative, goal-directed, self-disciplining, and reality accepting.

Finally, an evaluation of this synthesis by J. Bronowski, a leading natural scientist, has been included in Chapter 10.

This study leads to two important tentative conclusions. The first deals with what the five views hold in common. The contemporary humanistic psychologist views man primarily as a self-actualizing, self-determining being. That is to say, what distinguishes man as man is that his most central need-value is to create himself, through his own choices.

The second conclusion relates to how the views differ. The main distinction between the views of the five humanistic psychologists occurs with respect to what they postulate as the primary sources of guidance and energy for man's choices. Thus, Bonner thinks that man can be truly mindful of himself, fully self-actualizing, if he follows his natural yet socio-culturally conditioned proaction; Bugental emphasizes that man is urged toward authenticity by his feelingful awareness and not by any built-in instincts; Fromm considers man's reason as his most useful power for developing himself into a productive person; Maslow argues that man has a biologically built-in, though weak, instinct for synergic actualization; and May mentions man's powers of consciousness (feelingful awareness), conscience and courage as the three main powers that man has and must use to fulfill his one central need-value-task of self-determined self-actualization.

We turn now to a comparative appraisal of Bronowski's view of man. In his article, "A Twentieth Century Image of Man" Bronowski expresses the belief that man's primary distinguishing feature is that he creates imaginative plans and uses rational knowledge as the source of guidance for his choices.

Bronowski's definition of 'plans' is very broad, and incorporates both (1) man's strategies for creating himself (values) and (2) man's strategies for exploring, utilizing and creating phenomena external to himself. Bronowski's emphasis, however, is on the latter. This is in contrast to the views of the five humanistic psychologists who have emphasized the former. According to them what distinguishes man as man is that he is primarily focused on creating himself, on utilizing his powers to realize the higher human values and become the fully human person that he potentially is. As Fromm summarizes it,

While it is true that man's productiveness can create material things, works of art, and systems of thought, *by far the most important object of productiveness is man himself.*[1]

In conclusion, we need to examine the comments that Bronowski makes with respect to the view of man held by the five behavioral scientists. Bronowski notes that though they consistently stressed the human search for fulfillment within the self they neglected all that range of satisfactions which man finds in his exploration of nature. Rather than use the word 'neglected' however, it would be more correct to say that they gave a low degree of emphasis to man's need to explore and create phenomena other than himself and his fellow man. This comment of Bronowski's should be taken not as a criticism but as the clarifier of how his view differs from that of the behavioral scientists. Thus, while they have primarily emphasized that what distinguishes man as man is that he is a creator of himself into a synergic person, Bronowski has primarily emphasized that what distinguishes and has evolved man as man is that he is an explorer, utilizer and creator of the external environment.

This difference with regards to the primary distinguishing feature of man is of fundamental importance. Thus, Bronowski would like to see man make greater progress toward mastering and controlling external nature and for this reason he suggests rational knowledge as the most useful source of guidance for his choices. On the other hand, in direct contrast to this, the five behavioral scientists would like to see man make greater progress in the direction of actualizing his own humanness. They would have man transform himself into a synergic person, one who loves himself and his fellow man. One good example of a synergic person is Albert Schweitzer while Adolf Hitler is an example of an antisynergic person.

Man's main need-value-task is to create himself into a synergic person. To aid him in his choices toward full synergic

---

[1] Erich Fromm, *Man for Himself* (Greenwich, Conn.: Fawcett Publications, Inc., 1947), p. 97.

actualization, the five behavioral scientists (humanistic psychologists) have postulated certain primary sources of guidance. Thus, Bonner emphasizes proaction, Bugental points to feelingful awareness, Fromm to reason, Maslow to the instinctoid need for synergic actualization, and May to the powers of consciousness (feelingful awareness), conscience and courage. Instead of looking at their lower degree of emphasis on rational knowledge with a critical eye, we can appreciate their deeper understanding of the more basic sources of guidance that man must rely on in his joyous struggle to become a synergic person. Rather than call these sources irrational, it would be wiser to call them pro-rational or perhaps, even more accurately, superrational.

We see, therefore, that the primary distinguishing feature of man as man is that he creates himself through his own choices using superrational knowledge as his chief source of guidance and energy.

APPENDIX

PILOT STUDY

The pilot study has been used to design the content analysis instrument necessitated by the research study.

The steps involved in the design are the following: (*a*) Analyzing Hubert Bonner's book, *On Being Mindful of Man* so as to obtain the Implicit and Explicit Views of Man. The Explicit View of Man was obtained by directly noting relevant excerpts from the book. For the Implicit View of Man, statements relating to man's behavior were divided into two groups entitled: *Activities* and *Sentiments* respectively. This subdivision itself was prompted by George Homans' classification of human behavior into the three elements—activity, interaction, and sentiment. For the purposes of the present study, interaction was considered to be a special kind of activity and hence, only the two categories of activity and sentiment were used in the preliminary analysis required for the construction of the Implicit View of Man. (*b*) Deriving a pilot list of content analysis dimensions by means of a discriminating combination of the dimensions suggested by the analyses of the Implicit and Explicit Views of Man. (*c*) Adding dimensions to the pilot list so as to make the final list used in the content analysis instrument adequate both in terms of quality and logical correctness.

ANALYSIS OF PREFACE AND INTRODUCTION
—pp. xi-7

Activities
p. xi. Acts responsibly with respect to himself and others.
Oriented toward excellence and full actualization of his potential.

Sentiments
p. xi. Has a sense of duty toward himself and others; desires to perfect himself.

IMPLICIT VIEW OF MAN
p. xi. *Man works responsibly for full self-actualization; and, simultaneously for the self-actualization of others.*

EXPLICIT VIEW OF MAN
p. xi. Man is *unique, open,* and *creative; intentional* and *pro-active.*

Man is *a machine* and *an integral being; a reality* and *a potentiality; an ordinary creature* and *a superior being.*

Man is *responsible, self-actualizing, disciplined and self-controlled, aspiring and freedom-seeking.*

p. xii. Man is *consciously self-directive.*

p. 1. Man can influence his own evolution.

Man is a unique person, striving to actualize himself.

*****

Activities

p. 2. Man loves, has friendships, is courageous, and self-sacrificing.

Sentiments

p. 4. Men undergo anguish; alienation from themselves and the world; are concerned with life and death; have longings and disillusionments; experience loneliness and relatedness; yearn for expression and self-actualization.

p. 5. Man is his own maker — can make his own heaven and hell.

p. 6. Man experiences anxiety and guilt.

IMPLICIT VIEW OF MAN

Man experiences a wide range of sentiments.

Man can choose to have a life of joy or misery by choosing the values and ideals that he strives to realize in his life. *Man is self-creative.*

EXPLICIT VIEW OF MAN

p. 2. Man is *spontaneous.*

Man is *unpredictable.*

Man is authentically *forward-thrusting.*

Man is a *whole* being.

P. 7. The future dimension, in the form of forward-directedness, is more important than an individual's past.

The being of a person consists in his becoming.

Man changes as he moves in the direction of what he intends to become.

The true mark of a healthy man is a high degree of *self-inductiveness*.

Man dwells firmly in a foreworld.

The essence of humanness is the capacity to envision and actualize ideals and values.

Man chooses to create the life he aspires.

A man is what he will do.

ANALYSIS OF PART ONE : THE SUPREMACY OF SCIENCE – pp. 11-33

Activities

p. 21. Man strives for self-generated goals.

Man unceasingly makes choices.

p. 24. Man is the being who sees and feels, who is conscious of being conscious, who is unique and irreplaceable.

p. 27. Man reflects, consciously aspires, is creative and that is based on the pre-supposition that he has freedom of choice..

Sentiments

IMPLICIT VIEW OF MAN

Man *aspires and creatively reaches his goals.*

EXPLICIT VIEW OF MAN

p. 21. Man must relieve visceral tensions but even more powerfully, he must love, know and revere.

p. 27. Man's own experience is the evidence that he has *freedom of choice.*

p. 28. Man is the only being who is *conscious of himself*.

p. 29. Human beings, despite their uniqueness, are also *very much alike*.

ANALYSIS OF PART TWO : *REACTIVE MAN* — pp. 37-57
EXPLICIT VIEW OF MAN

p. 46. *Idealism* is alive in the heart of man.

Man creates his own destiny.

p. 51. Man is characterized by *purpose*, by changes in goals, by proaction.

p. 57. Man aspires, transforms himself, through love, guilt, suffering tragedy, and his creative advance into the future.

ANALYSIS OF PART THREE: *PROACTIVE MAN* — pp. 61-199
Activities

p. 69. Our choice today is quite literally between an agonizing peace and mutual extermination.

Sentiments

p. 61. Man suffers libidinal and instinctual conflicts but also, value and meaning conflicts. Man has sexual, marital, vocational problems but equally, if not more, he has spiritual problems.

p. 63. Man is dissociated from nature, from himself, and from other men.

IMPLICIT VIEW OF MAN

Modern man is faced with many problems and they join together in producing the alienation and annihilation of individual personality.

EXPLICIT VIEW OF MAN

p. 63. Despite the horrendous crises that have afflicted civilization, humanity is fundamentally proactive and *self-renewing*.

Man is both producer and product of the human situation.

p. 65. The essence of modern man's predicament is a profoundly moral one.

p. 70. Only a psychology that endows man with the capacity for self-transformation, is a psychology that is fit for the needs of man in the mid-twentieth century.

Human compassion is the last hope of conserving our civilization.

*****

Activities

p. 71. Man's constructive efforts, guided by the vision of a long-term ethics, have been without exception magnificent forward leaps. Exhausted by his savageries, man has nevertheless comprehended his own condition and, through compassion and other-mindedness, built himself into a new life.

Sentiments

IMPLICIT VIEW OF MAN

Man has the capacity to create and to love over and above his capacity to destroy and to hate.

EXPLICIT VIEW OF MAN

p. 70. Civilization must rely wholly on the long-term ethics which is functionally and ontologically bound up with man's capacity to love.

p. 71. In the face of an unimaginable catastrophe, reliance on *man's creative and moral character* is the only practical alternative.

*****

Activities

pp. 72—78. Modern man is paralyzed by an inaccurate view of man.

Modern man has ceased to think; he has become into a robot-like parrot.

Through disuse of his intellect, he has contributed to his depersonalization.

The pain of estrangement can be mollified by the enhancement of human relatedness.

The social situation of our time is one of conflicting values, a mood of pessimism.

...In this situation alienation and estrangement on a wide scale are inescapable.

Sentiments

p. 73. Modern man is gripped by fears and anxieties; by a feeling of helplessness and futility; by a paralysis born out of a view of man as a helpless victim of forces over which he has no control.

p. 74. The possibility of death makes man acutely aware of his finitude. Reverence for life does not preclude a poetic regard for the dignity of death.

IMPLICIT VIEW OF MAN

Modern man is paralyzed and fearful because he has internalized an inaccurate view of his nature; rather, than see himself as an *active creator* he views himself as passive and determined. Thus, his failure to strive courageously to realize his potentialities makes him feel impotent.

Death is part of nature. It is better to realistically accept its inevitability and to then decide to live more fully while you do have the opportunity.

Man's alienation from man is a product of depersonalizing, impersonal, social conditions and philosophies. Man can overcome this and other psychological ills through the full use of his powers to think, to trust, to lovingly commit himself to care for his fellow human beings.

p. 72. Modern man is moved far less by his capacity to choose than by his awareness of his own finitude.

Today, man has no sure anchorage, no personal basis of values. He is disoriented, his work is meaningless for him, and he has no sense of direction. This condition, not the inherent degeneracy of a people, is the source of the diabolic behavior of men.

Constructive behavior is based on the self-confidence born of the awareness of one's finitude and one's potentiality for transforming intentions into actions.

p. 74. Anguish, anxiety and despair are ontological conditions. They are immanent in human nature.

p. 75. Although man's fear of death may cause serious emotional disturbances, it is a man's fear of life, reality, that is the source of most of our psychological ills.

*****

Activities

p. 79. We are guilty because we can choose. Guilt, like choice, is an attribute of our human nature, a potentiality of human life.

p. 81. Human beings share a common burden, the responsibility of sealing their fate in the act of choice.

Because in his choosing man chooses for other human beings like himself, the problem of freedom turns out to be a cognitive than a moral one.

Sentiments

p. 79. Guilt is not only a feeling of moral failure. It is not merely a form of psychological crippling, but a disordered state of being.

p. 81. Man's helplessness in the face of uncontrollable events is the consequence, not of a lack of freedom but because of it.

IMPLICIT VIEW OF MAN

Man has the moral responsibility to make creative, intelligent self-actualizing choices. *By choosing carefully, man can overcome his feelings of impotence.*

Real guilt is a part of the nature of man, and follows from his freedom of choice.

EXPLICIT VIEW OF MAN

p. 79. Man is guilty when he compromises truth, when he does not combat injustice, when he submits to evil, when he fails to actualize his potentialities.

p. 80. All psychologists will soon need to reckon with the *morality* and *spirituality* which, though not overtly identifiable, are the prime movers of man's being.

p. 82. The fully human quality of man's inner life is not unconscious but *conscious*. It consists in the capacity and the act of choice.

Man is a decision-making being who, in choosing, is faced with the terrifying possibility that the outcome of his choice is highly problematic.

*****

Activities

p. 81. The history of man is the record of his decisions and all their creative and destructive consequences. Every choice is based on a scale of values and has consequences for others. Even the decision not to choose, is an act of choice.

p. 85. All living stuff acts toward the consummation of a purpose.

p. 88. Moral behavior is thus one of the highest, if not the supreme, manifestations of volition, deliberation, and choice.

p. 89. The freedom of choice affirms the proactive nature of man.

p. 98. No man in possession of his powers can escape accountability for his acts. He alone among animals, has the dreadful responsibility of choosing between good and evil, and he alone is conscious of possessing it.

Sentiments

p. 84. Our profoundest, our most significant, forms of conduct presuppose the freedom of choice. Guilt, regret, responsibility, and anxiety all imply a measure of choice.

p. 90. At no other period in history has man been more confronted with the inescapable fate of all men : the fact that all men must choose.

p. 95. Neither the antecedents nor the consequents of human choice are purely intellectual. Their vitality and urgency lie in their affective nature. They are basically feelings, sentiments, and passions. Choice is seldom motivated by coldly intellectual considerations.

IMPLICIT VIEW OF MAN

Man is ceaselessly confronted with the responsibility to make choices. Healthy choices are accompanied by real feelings of anxiety, etc.; neurotic choices are accompanied by neurotic feelings of anxiety, guilt, etc.

Man's essential nature is simply described thus: *he sets goals for himself and then accomplishes them.*

EXPLICIT VIEW OF MAN

p. 82. The centre of man's orientation has shifted from Descartes' *Cogito* to *Jaspers' ich wahle* — "I choose."

p. 84. Wisdom, if it means anything psychologically, is the recognition by each of us, that each man is responsible for his own destiny.

p. 85. The mature man is the organized totality of self-regulations, of goal-directedness, of intentions in the process of actualization.

*****

Activities

p. 101. Human behavior is rooted in moral and philosophical beliefs.

p. 104. Freedom is not an abstraction but an act; we can remain free only by actualizing our freedom through continuous practice.

p. 108. Freedom to choose, always incurs responsibility toward others not only oneself.

Sentiments

p. 104. Man feels most free and independent when no matter what the obstacles may be, he can express his need for privacy, individuality, and autonomy.

IMPLICIT VIEW OF MAN

Man has the freedom and the responsibility to make his own choices and thus shape his own psychophysical destiny.

Man behaves according to his philosophical beliefs.

Man obtains a sense of freedom by striving responsibly for self-actualization.

EXPLICIT VIEW OF MAN

p. 86. Man differs from all other animals in his *capacity to choose*.

p. 91. The basic attitudes of *commitment and responsibility* are both reflections of man at his highest and best. Both of them rest on man's capacity to make choices.

p. 97. Man's biological make-up limits his freedom of choice. Psychological conflict limits our capacity to choose.

Man's image of himself aids or influences his freedom of choice.

Lack of knowledge is an impediment to free choice.

The social situation, particularly the cultural barriers affect our freedom of choice.

p. 98. The healthy individual makes commitments and decisions.

p. 103. Significant human acts, acts which make a profound difference in human affairs, are acts of the human spirit.

Man is first, not last, a self-affirming being. The center of his existence is neither nature, nor society but himself.

Man proves his existence, or validates his individuality, by means of the courage to be himself.

Freedom and the need to know (manifested in curiosity and the search for meaning) are powerful human needs.

Only that man is free who lives in a society of free men.

*****

Activities

p. 118. We strive for things ahead of us and are not driven solely by events behind us.

p. 119. The sick person, in his fear of being himself, of actualizing his potentialities distorts his potential uniqueness beyond recognition. He is sick because his self-activity is weak or atrophied.

p. 121. Life is the creative advance into the future.

Sentiments

P. 117. Anxiety is a sign that we care, that persons and events are important to us.

Because he lives in an open, unstructured world, the proactivert is subject to more anxiety, but the very openness of his being permits him to absorb his anxiety in a constructive movement toward the future.

p. 119. Without the future there can be no present. In the stream of psychological time the future is particularly important.

IMPLICIT VIEW OF MAN

Proactive man is *future-oriented*, and adventurous. He accepts anxiety as a natural accompaniment of his striving for self-created goals.

Man's goal-creating and goal-accomplishing nature assures his psychological health.

EXPLICIT VIEW OF MAN

p. 109. Man cries out for a world that will provide the conditions for self-actualization.

Man cannot perform the feat of self-transformation in utter isolation. It always takes place in the context of a living moral order.

The proactive individual lends himself freely in the service of others.

The reality of the self is validated by its participation in the world of other selves.

The individual who labors toward self-development eventually transforms society.

Man's most desperate longing is not group-relatedness but self-fulfillment.

Man is truly his own foundation.

p. 117 Nothing is more characteristic of the proactive individual than *openness* to experience.

p. 120. *Self-activity,* or becoming, is the authentic sign of psychological health, of movement toward self-perfection.

Activities

p. 125 Human life is replete with examples of men who set out to search fo one thing only to find something quite different.

p. 129 Human life is the pursuit of the possible, and the possible lies in the future. There is nothing mystical or mysterious in this forward directedness; it contains no idea of man's 'destiny', but of his opportunity. Destiny implies a push fom behind . . . rather than a future of unlimited opportunities.

Sentiments

p. 125. One's past can hold no surprises for it is already completed; one's present is too immediate to cause more than a fleeting emotion, but the future can be full of astonishment and wonder.

IMPLICIT VIEW OF MAN

If man *seeks courageously,* he shall receive abundantly. Man's essential nature is to courageously strive for self-created objectives and goals, and simultaneously be *open to experience and feedback.*

EXPLICIT VIEW OF MAN

p. 121. The healthy creative individual lives in the process of becoming.

Man is a seeker of future ends.

The sick individual regrets his past, abhors his present, and dreads his future.

p. 124. The highest type of human being feels responsible for the welfare of humanity.

p. 125. Without adventure into the uncharted future, civilization sinks back into decay.

Man is a dreamer of dreams.

The proactive individual knows that, with effort, he can realize perfection by successive approximations.

p. 128. Man is a self-directing, freely choosing, value-creating individual.

p. 129. Every man, however much he is like other men, is unique and irreplaceable.

p. 130. In the end, the act of fulfilling himself as a person is the only thing worth having.

*****

Activities

p. 131. The human person is in a constant process of becoming.

p. 132. What gives every person his being, his personality, is the person himself.

p. 133. The neurotic person is profoundly deficient in the process of becoming an authentic person, of actualizing his personality, of self-acceptance, and of entering into the self of another.

p. 137. The neurotic individual immobilizes himself through the compulsion to appear different from what he is.

Sentiments

p. 132. Basic to all achievements is man's profound dissatisfaction with things as they are.

p. 135. Creative tensions are the catalysts of his forward movement.

p. 139. There is something sad in the bondage to the past and the constant leaning upon the person of another for one's fulfillment, in the failure to actualize one's inner strengths, the strengths which are realized in the process of adventure and in the encounter with becoming.

IMPLICIT VIEW OF MAN

The healthy person accepts himself and courageously strives to make himself into a more healthy person. i.e., a more loving and more productive person, in short, a more effective person.

EXPLICIT VIEW OF MAN

p. 131. Man's being consists in his becoming.

p. 132. *Man creates his own personality,* his own reality.

p. 135. The healthy, proactive individual copes productively with the world. Being motivated by ideals as well as hungers and deficiencies, the healthy individual is perpetually "on the go."

p. 136. Healthy existence is not a comfortable state of homeostasis or inertia, but of striving and becoming.

Growth is no mere unfoldment of inherent traits, but a creative transformation.

p. 140. The primary striving of the individual is toward *self-affirmation and self-validation.*

*****

Activities

p. 140. The impulsion toward health is the striving to become oneself. Psychological sickness consists in the blockage of the process of becoming oneself.

p. 157. Love is the fusion of one's whole nature in the nature of the other person.

p. 164. Love moves men and impels men to move mountains and so moves life on to greater perfectibility.

Sentiments

p. 142. Guilt results from the failure to actualize one's potentialities.

p. 143. The person who *cares* for the becoming of another person is himself swept along by the other's self-actualization.

p. 150. Self-love, then, is not selfishness but that high respect for oneself which is the mark of the healthy, self-confident, and loving individual.

p. 157. Love like the rest of life is a process. . .

p. 158. Love is a human creation.

p. 159. Love as a creative encounter is the expression of care, responsibility, faith, and sacrifice.

p. 163. Men love, sacrifice, revere. . .

p. 164. Love, care, sacrifice and faith are man's most treasured sentiments.

IMPLICIT VIEW OF MAN

If man does not strivingly aspire to become his true self, he becomes sick and neurotic.

Through loving himself, the Thou and the relationship, man affirms himself and is motivated to become all he can.

EXPLICIT VIEW OF MAN

p. 147. Self-actualization, the realization of one's inner life, can be achieved only through communication with another.

p. 157. The surest way not to achieve happiness whether in love or in the whole of life, is to consciously seek it.

p. 167. Every healthy person strives for greater individuation and self-actualization.

Man is an open personality; his future, while he lives, is never closed.

An individual is not a passive respondent to stimuli but is a self-transforming person.

*****

Activities

p. 165. Personality is a way of striving toward self-perfection, and this striving is the individual's style of life. The style of life is never fully made, but always in the making. The modes of striving are as numerous as the individual's themselves.

p. 167. Every behavior of the total person bears the stamp of his own individuality, his own style of living.

Sentiments

p. 171. Life is an adventure full of risks as well as satisfactions.

p. 172. The style of life is the set of ideals which a person selects from many possibilities, and to which he pledges his loyalty, above all other sentiments. In this way, man conforms freely to one's expectation: his image of himself.

IMPLICIT VIEW OF MAN

Man creates his own personality, his style of life so as to optimally accomplish his life-goals, to become what he potentially is or can become.

EXPLICIT VIEW OF MAN

p. 167. By means of our style of life each of us can make of his life a work of art.

A prime condition of effective and self-enhancing conduct is the self-corrective quality of the individual's way of life.

A style of life, like life itself, is more than a guide to conduct; it is the art of creating one's own life.

p. 179. Proactive life, i.e., *the truly human life, is an endlees becoming.* It is an endless seeking of a perfection that can never be attained. The proactive man is never certain of the outcome of his search. *He feels unfinished and never wholly fulfilled.*

Man is a being constantly engaged in actualizing himself by means of his individual life style.

Man's life style is his philosophy of living; his philosophy in action.

*****

Activities

p. 182. Proactive man is a being who is lured by adventure and risk, by improvisation and conceptual experimentation.

p. 183. The healthy man is self-propulsive.

Psychological growth—which is to say psychological health—is the courage to face the truth about ourselves and the daring spirit of actualizing our authentic lives in the face of criticism, rejection and the threat of isolation.

Sentiments

p. 181. Man must be viewed as a life-totality.

p. 182. Proactive man rejects the all importance of security. . .

p. 183. Psychological illness is in large measure the fear by the individual of facing the truth about himself.

Authentic human relationships are embedded in honest self-acceptance. The more a person accepts himself, the more open he is to the selves of other persons.

IMPLICIT VIEW OF MAN

Proactive man is simultaneously an integrated and an integrating man—an *integrative* man.

Proactive man accepts himself and *courageously strives forward to actualize his authentic self.*

EXPLICIT VIEW OF MAN

p. 181. Man is a life-totality and must be viewed as a holistic being.

Human beings are inherently both reactive and proactive. Proaction varies, like intelligence or artistic ability, on a continuum from most to least. The proactive individual is characterized by a high degree of forward-directedness, independence and becoming.

p. 182. Every individual has the capacity to resist the molding force of his social environment and become what he potentially is.

p. 183. *The healthy man is a daring and challenging individual.*

p. 184. Self-affirmation is our most despairing need.

p. 192. The conscious self is the organizing and self-regulating capacities of the total personality.

p. 193. Personality is a unique whole.

p. 196. The self is always discovered in association with others — in love and strife and reconciliation.

p. 198. The most important fact of human life is that man is a being who, while resting on his past, builds himself a future.

Personality is the "organ" of self-direction and self-actualization.

ANALYSIS OF PART FOUR : TOWARD THE PERFECTIBILITY OF MAN and epilogue — pp. 203-234

EXPLICIT VIEW OF MAN

p. 226. Man is a creative and proactive being.

Self-transformation is the most difficult of all tasks.

Eash person is responsible for his own development.

Man is the being that strives for the maximum realization of his values and ideals.

The proative man bestows his own finality upon himself.

Each person is unique and irreplaceable.

Man is free to actualize his potentialities.

p. 233. Man is the maker of his own life.

p. 234. *A total knowledge of the whole man is impossible.* Man is the being who reveres himself; has reverence for himself.

SOME POSSIBLE DIMENSIONS FOR A CONTENT ANALYSIS INSTRUMENT

The following dimensions have been obtained from an analysis of the Implicit and Explicit Views of Man outlined in the previous pages. Special attention was given to the italicized words or lines in making the selection.

POSSIBLE DIMENSIONS SUGGESTED BY AN ANALYSIS OF THE EXPLICIT VIEW OF MAN

1. Unique
2. Open.
3. Creative.
4. Intentional/Purposive/Goal-Directed.
5. Proactive/Forward-Thrusting.
6. An integral being/An organized totality.
7. A potentiality.
8. Consciously self-directing.
9. Responsible.
10. Self-actualizing.
11. Disciplined and self-controlled; self-regulating.
12. Aspiring/Idealistic/A dreamer.
13. Freedom-seeking.
14. Spontaneous.
15. Unpredictable/Capable of making choices.
16. Conscious of self.
17. Much like fellow man.
18. Self-renewing.
19. Aware of his finitude.
20. Anguished/Anxious/Despairing.
21. Moral/Spiritual.
22. A conscious decision-maker.
23. A social being.
24. Capable of making commitments.
25. Self-affirming/Self-validating.
26. Self-transforming/Self-creative.
27. Other-minded/Other-affirming.
28. Self-fulfilment oriented.
29. Self-active.
30. Future-oriented.
31. Always in process.

32. Always unfinished.
33. Reactive.
34. Daring and challenging.
35. Ultimately unknowable.
36. Self-respecting.

POSSIBLE DIMENSIONS SUGGESTED BY AN ANALYSIS OF THE
IMPLICIT VIEW OF MAN

1. Responsible for self-actualizing/Ontologically responsible.
2. Responsible for the self-actualizing of others/Socially responsible.
3. Capable of experiencing a wide range of sentiment.
4. Self-creative.
5. A goal-setting and accomplishing being.
6. Capable of loving over and above his capacity to hate.
7. An active creator.
8. Capable of free choice.
9. Liable to experiencing real guilt.
10. Capable of overcoming his feelings of impotence.
11. Ceaselessly confronted with choices.
12. Thou-loving.
13. Future-oriented.
14. Adventurous.
15. Accepting of anxiety.
16. A courageous striver.
17. Open to experience and feedback.
18. Self-accepting.
19. Self-developing.
20. An integrative being.
21. Self-actualizing.

PILOT LIST OF CONTENT ANALYSIS DIMENSIONS

The following list of twenty-two dimensions is a discriminating combination from the possible dimensions suggested by the analyses of the Explicit and Implicit Views of Man.

Given below are the serial numbers of the dimensions selected from analyses of :

		Explicit View of Man	**Implicit View of Man**
1.	Unique	1	—
2.	Open to life/experience	2	17, 20
3.	Creative	3	7
4.	Goal-directed	4	5
5.	Forward-thrusting	5	
6.	Holistic/An integrated whole	6	20
7.	Self-determining	8, 22	
8.	Ontologically responsible	9, 24	1
9.	Socially responsible	9, 23, 24	2
10.	Self-actualizing/Self-realizing	10, 18, 25, 26, 28, 32	4, 19, 20, 21
11.	Self-disciplining	11, 33	
12.	Future-imagining	12, 30	13
13.	Other-affirming	27, 28	2, 12
14.	Freedom-cherishing	13, 15	8
15.	Conscious/Aware	7, 8, 16, 19, 22	
16.	Much like fellow man	17	
17.	Accepting of non-hedonic emotions	20	3, 9, 15
18.	Self-affirming	14, 25, 28, 36	18
19.	Self-active	29	7
20.	Always in process	31	
21.	Risk-taking	34	14, 16
22.	Ultimately unknowable	35	

1. Rational

In an article in *The Humanist* (March-April, 1971) H.J. Eysenck writes, ". . .at the bottom of humanist attitudes lies belief in the power and importance of reason."

2. Intelligent

Concluding his article, "Man's Place in the Physical Universe" (in the book, *New Views of the Nature of Man)* Willard F. Libby notes, ". . . This, to me, is man's place in the physical universe: to be its king through the power he alone possesses — the Principle of Intelligence."

3. Present-confronting

In an article in the *Journal of Humanistic Psychology* (XI, no. 1, Spring 1971) James F.T. Bugental explains the humanistic ethic. One of the points he makes is that, "one always lives only at the present moment. . . it is only in the ever-flowing present that we can realize our own potentials." This here and now perspective is taken into account in this dimension.

4. Nature-appreciating

In his book *The Philosophy of Humanism* Corliss Lamont refers to the appreciation of nature's beauty and loveliness as one of the central propositions of the Humanist Philosophy. (p. 13)

5. Sexual

Robert S. de Ropp in his book, *Sex Energy* writes, ". . . If the mind of contemporary man seems obsessed by this force let us admit that there is a reason for the obsession. For nature herself is sex-obsessed. If there is one mechanism we can be almost certain of finding in any form of life we choose to study it is the sexual mechanism." (p. 3)

6. Pleasure-loving

In this book entitled, *Pleasure* Alexander Lowen notes, ". . . Bodily pleasure is the source from which all our good feelings and good thinking stems. If the bodily pleasure of an individual is destroyed, he becomes an angry, frustrated, and hateful person." (p. 11) "Pleasure is the creative force in life." Lowen distinguishes between so-called 'fun' and the experience of pleasure. He writes, "The search for fun by adults undermines their capacity for pleasure. Pleasure demands a serious attitude toward life, a commitment to one's existence and work. It may be viewed as the business of life, whether one is playing as a child or working as an adult. An escapade, regardless of its apparent fun, must end in pain, as do all attempts to escape the commitment to life." (p. 18)

7. A communicating being

This dimension has been added on the basis of my own experience on the assumption that what is true of me—my deep need to communicate with other people — is also true of man in general.

8. Other/Miscellaneous

This dimension/category is being kept in the content analysis instrument in case the content analysis of the selected works reveals one or more dimensions that cannot be logically accommodated under the remaining dimensional categories.

THE CONTENT ANALYSIS INSTRUMENT

(a) The dimensions can be understood as a series of predicates for the sentence beginning, "Man is . . . . . ."

(b) The dimensions have been listed in alphabetical order apart from the Other/Miscellaneous category which is given at the end.

No.	Dimension/Category	Count	Total	Percentage Degree of Emphasis
1.	A communicative being			
2.	Accepting of non-hedonic emotions			
3.	Always in process			
4.	Conscious/Aware			
5.	Creative			
6.	Forward-thrusting			
7.	Freedom-cherishing			
8.	Future-imagining			
9.	Goal-directed			
10.	Holistic/An integrated whole			
11.	Intelligent			
12.	Much like fellow man			
13.	Nature-appreciating			
14.	Ontologically responsible			
15.	Open to life/experience			

16.	Other-affirming			
17.	Pleasure-loving			
18.	Present-confronting			
19.	Rational			
20.	Risk-taking/Courageous			
21.	Self-active			
22.	Self-actualizing/Self-realizing			
23.	Self-affirming			
24.	Self-determining			
25.	Self-disciplining			
26.	Sexual			
27.	Socially responsible			
28.	Ultimately unknowable			
29.	Unique			
30.	Other/Miscellaneous			
			Cumulative Total	**= 100%**

BIBLIOGRAPHY

A. BOOKS

Adler, Alfred. *The Science of Living.* New York : Anchor Books, 1969.

—. *Understanding Human Nature.* Greenwich, Conn. : Fawcett Publications, Inc., 1954.

Allport, Gordon W. *The Person in Psychology.* Boston : Beacon Press, 1968.

—. *Becoming.* New Haven : Yale University Press, Inc., 1955.

Arnspiger, V. Clyde, W. Ray Rucker, and Mary E. Preas. *Personality in Social Process.* Dubuque, Iowa : Wm. C. Brown Book Company, 1969.

Assagioli, Roberto. *Psychosynthesis.* New York : The Viking Press, 1965.

Baller, Warren R. (ed.). *Readings in the Psychology of Human Growth and Development.* 2d ed. New York : Holt, Rinehart and Winston, Inc., 1969.

Barrett, William. *Irrational Man.* New York : Anchor Books, 1958.

Berelson, Bernard, and Gary A. Steiner. *Human Behavior.* New York : Harcourt, Brace and World, Inc., 1967.

Bonner, Hubert. *On Being Mindful of Man.* Boston : Houghton Mifflin Company, 1965.

Bronowski, J. *The Identity of Man.* Revised edition. New York : American Museum Science Books, 1966.

—. *Science and Human Values.* Revised edition. New York : Harper Torchbooks, 1965.

—. *The Common Sense of Science.* New York : Vintage Books.

Buber, Martin. *Between Man and Man.* New York : The Macmillan Company, 1965.

—. *I and Thou.* New York : Charles Scribner's Sons, 1958.

Bucke, Richard Maurice. *Cosmic Consciousness.* New York : E.P. Dutton and Company, Inc., 1923.

Bugental, James F.T. *The Search for Authenticity.* New York : Holt, Rinehart and Winston, 1965.

—. (ed.). *Challenges of Humanistic Psychology.* New York : Mc-Graw-Hill Book Company, 1967.

Cassirer, Ernst. *An Essay on Man.* New Haven : Yale University Press, 1944.

Chapple, Eliot D. *Culture and Biological Man.* New York : Holt, Rinehart and Winston, Inc, 1970.

Chiang Hung-Min and Abraham Maslow. *The Healthy*

Personality. New York : Van Nostrand Reinhold Company, 1969.

Cox, David. *Modern Psychology.* New York : Barnes and Noble, Inc., 1968.

De Chardin, Teilhard. *The Phenomenon of Man.* New York : Harper Torchbooks, 1959.

De Ropp, Robert S. *Sex Energy.* New York : Delta Books, 1969.

Dewey, John. *Human Nature and Conduct.* New York : The Modern Library, 1957.

Festinger, Leon, and Daniel Katz. (eds.). *Research Methods in the Behavioral Sciences.* New York : Holt, Rinehart and Winston, 1953.

Fromm, Erich. *The Revolution of Hope.* New York : Bantam Books, 1968.

—. *The Heart of Man.* New York : Harper Colophon Books, 1964.

—. *Beyond the Chains of Illusion.* New York : Pocket Books, Inc., 1962.

—. *Man for Himself.* Greenwich, Conn. : Fawcett Publications, Inc., 1947.

Fromm, Erich and Ramon Xirau. (eds.). *The Nature of Man.* New York : The Macmillian Company, 1968.

Giorgi, Amedeo, *Psychology as a Human Science.* New York : Harper and Row, Publishers, Inc., 1970.

Goldstein, Kurt. *Human Nature.* New York : Schocken Books, 1963.

Homans, George C. *The Human Group.* New York : Harcourt, Brace and World, Inc., 1950.

Howe, Reuel L. *The Miracle of Dialogue.* New York : The Seabury Press, Inc., 1963.

Jung, C.J. *The Undiscovered Self.* New York : Mentor Books, 1958.

Lamont, Corliss. *The Philosophy of Humanism.* 5th ed. New York : Frederick Ungar Publishing Company, 1965.

Lowen, Alexander. *Pleasure.* New York : Lancer Books, 1970.

Maslow, Abraham H. *Motivation and Personality.* 2d ed. New York : Harper and Row, 1970.

—. *Religions, Values, and Peak-Experiences.* New York : The Viking Press, 1970.

—. *Toward a Psychology of Being.* New York : Van Nostrand Reinhold Company, 1968.

—. *The Psychology of Science.* Chicago : Henry Regnery Company, 1966.

—. (ed.) *New Knowledge in Human Values.* Chicago: Henry Regnery Company, 1959.

Matson, Floyd W. *The Broken Image.* New York : Anchor Books, 1964.

—. (ed). *Being, Becoming and Behavior.* New York : George Braziller, Inc., 1967.

—. and Ashley Montagu (eds.). *The Human Dialogue.* New York : The Free Press, 1967.

May, Rollo. *Existential Psychology.* New York : Random House Inc., 1969.

—. *Psychology and the Human Dilemma.* Princeton, New Jersey : D. Van Nostrand Company, Inc., 1967.

—. *Man's Search for Himself.* New York : Signet Books, 1953,

McDermott, Robert A. (ed.) *Radhakrishnan.* New York : E.P. Dutton and Company, Inc., 1970.

Montagu, Ashley. *The Direction of Human Development.* New York : Hawthorn Books, Inc., 1970.

—. *The Humanization of Man.* New York : Grove Press, Inc., 1962.

—. *Man in Process.* New York : Mentor Books, 1961.

Osborn, Arthur W. *The Expansion of Awareness.* Wheaton, Illinois : Quest Books, 1961.

Otto, Herbert A. (ed.). *Human Potentialities.* St. Louis, Missouri : Warren Green, Inc., 1968.

Ouspensky, P.D. *The Fourth Way.* New York : Vintage Books, 1957.

—. *In Search of the Miraculous.* New York : Harcourt, Brace and World, Inc., 1949.

Peterson, Severin. *A Catalog of the Ways People Grow.* New York : Ballantine Books, 1971.

Platt, John R. (ed.). *New Views of the Nature of Man.* Chicago: The University of Chicago Press, 1965.

Polanyi, Michael. *Personal Knowledge.* New York : Harper Torchbooks, 1962.

Rogers, Carl R. *On Becoming a Person.* Boston : Houghton Mifflin Company, 1961.

— and Barry Stevens, *Person to Person : The Problem of Being Human.* New York : Pocket Books, 1967.

Rogers, Raymond. *Coming into Existence.* New York : Delta Books, 1967.

Ruitenbeek, Hendrik M. (ed.). *Varieties of Personality Theory.* New

York : E.P. Dutton and Company, 1964.

Selye, Hans. *From Dream to Discovery.* New York : McGraw-Hill Book Company, 1964.

Severin, Frank T. (ed.). *Humanistic Viewpoints in Psychology.* New York : McGraw-Hill Book Company, 1965.

Skinner, B.F. *Beyond Freedom and Dignity.* New York : Alfred A. Knopf, 1971.

Sutich, Anthony J., and Miles A. Vich. (eds.). *Readings in Humanistic Psychology.* New York : The Free Press, 1969.

Tillich, Paul. *Dynamics of Faith.* New York : Harper Torchbooks, 1957.

—. *The Courage to Be.* New Haven : Yale University Press, 1952.

Warner, Samuel J. *Self-Realization and Self Defeat.* New York : Grove Press, Inc., 1966.

Watts, Alan W. *The Book.* New York : Collier Books, 1966.

—. *Nature, Man and Woman.* New York : Vintage Books, 1958.

Wilde, Jean T., and William Kimmel. *The Search for Being.* New York : The Noonday Press, 1962.

B. ARTICLES

Bugental, James F.T. "The Humanistic Ethic — The Individual in Psychotherapy as a Societal Change Agent," *The Journal of Humanistic Psychology,* XI (Spring, 1971), 11-25.

Eysenck, H.J. "Reason with Compassion," *The Humanist* (March-April, 1971), pp. 24-25.

Harman, Willis W. "The New Copernican Revolution." *The Journal of Humanistic Psychology,* IX (Fall, 1969), 127-134.

C. UNPUBLISHED WORKS

Charron, William Cletus. "An Exposition and Analysis of William James's Views on the Nature of Man." Unpublished Doctoral Dissertation, Marquette University, 1966.

Hater, Robert James. "Psycho-Philosophy of the Human Person: (Gordon W. Allport's Dynamic Theory of Human Personality, with Special Emphasis on the Philosophical Implications Derived from the Study of his Approach to Man)." Unpublished Doctoral Dissertation, St. John's University, 1967.

www.ingramcontent.com/pod-product-compliance
Lightning Source LLC
LaVergne TN
LVHW052257210726
843527LV00040B/549